Essays on geometric group theory

Ramanujan Mathematical Society Lectures Notes Series

Vol. 1: Number theory
Vol. 2: The Riemann zeta function and related themes
Vol. 3: Formal language aspects of natural computing
Vol. 4: Commutative algebra and combinatorics
Vol. 5: Convexity in discrete structures
Vol. 6: Number theory and discrete geometry
Vol. 7: Discrete mathematics
Vol. 8: Lectures on operator theory
Vol. 9: Essays on geometric group theory
Vol. 10: Techmüller theory and moduli problems

Ramanujan Mathematical Society

Lecture Notes Series

Volume 9

Essays on geometric group theory

Proceedings of the instructional workshop and international conference on group theory, Indian Institute of Technology, Guwahati, December 2002

Volume editor

N. S. Narasimha Sastry

International Press
www.intlpress.com

Ramanujan Mathematical Society
Lecture Notes Series, Volume 9
Essays on geometric group theory

Volume Editor:
N. S. Narasimha Sastry

Published in 2010 by International Press, Somerville, Massachusetts, U.S.A. under license from the Ramanujan Mathematical Society.

This work is also published in the Republic of India, exclusively by the Ramanujan Mathematical Society.

ISBN 978-1-57146-194-0

Typeset using the LaTeX system.
Printed in the United States of America.

Preface

Geometric ideas have played a very major role in the development of Group theory in 20th century. Coxeter groups and Coxeter complexes, Tits theory of buildings, Bass-Serre theory of Groups acting on trees, Gromov's notion of hyperbolic groups and combinatorial methods have all been very central in several areas of mathematics.

With the objective of introducing the basic ideas and recent developments in these topics to young researchers in various Research Institutions and Universities in India, the Indian Statistical Institute at its Bangalore Centre and the Indian Institute of Technology, Guwahati, organized "Instructional Workshop and International Conference on Geometric Group Theory" during December 2–21, 2002, at the beautiful campus of the Indian Institute of Technology, Guwahati.

In the Workshop part of this programme (December 2–14, 2002), five short courses of lectures were given. The topics were:

- Coxeter Groups (10 lectures) by R.P. Shukla and Maneesh Thakur
- Buildings and Tits Systems (10 lectures) by N.S. Narasimha Sastry
- Bass-Serre Theory (10 lectures) by A. Raghuram and B. Sury
- Hyperbolic Groups (5 lectures) by A. Naolekar
- Combinatorial Group Theory (5 lectures) by I.B.S. Passi and Roman Mikhailov

In the Conference part of this programme (December 16–21, 2002), experts in this area of mathematics were invited to give research/expository lectures in the topic of their expertise. The articles in this volume are mostly the expanded versions of the lectures delivered by the authors in this Conference. It is hoped that this volume will be of some assistance to mathematicians who would like to familiarize themselves with these topics.

The Organizing Committee for this programme consisted of Professors I.B.S. Passi (Chairman), N.S. Narasimha Sastry, B. Sury, B.K. Sarma and Meenaxi Bhattacharjee.

Apart from the Indian Statistical Institute and the Indian Institute of Technology, Guwahati, the programme also received financial support from the National Board for Higher Mathematics, India; The British Council, India; Harish-Chandra Research Institute, Allahabad; Oil and Natural Gas Commission and Oil India Limited.

I wish to thank all the speakers in this programme and the authors of the articles, without whose enthusiastic support, this programme would not have achieved its objective.

As a member of the organizing committee, I also thank all the sponsors for their financial support; the Indian Institute of Technology, Guwahati, for the excellent facilities and enthusiastic and cheerful support extended during the Conference; and the Ramanujan Mathematical Society for bringing out this volume.

I also would like to record my appreciation of the enthusiasm, meticulous attention to detail and the caring of Professor Meenaxi Bhattacharjee, her colleagues and students at IIT, Guwahati, which contributed largely towards making this event a very pleasant and fruitful experience. I am confident that all the participants at this event share this sentiment.

Finally, I would like to thank Ms. Asha Lata for her help in compiling this volume.

Date: 5 December 2008
Indian Statistical Institute
Bangalore Centre

N. S. Narasimha Sastry
Editor

Talks given in the Conference

Professor Peter Abramenko	Some applications of Bass-Serre theory to the GE_2 question
Professor Roger M. Bryant	Metanilpotent varieties of groups
Professor Marc Burger	Tree automorphisms, products of trees and lattices
Professor Alan Camina	Block designs
Dr. Indira Chatterji	Twisted idempotent conjectures
Dr. Ahmad Erfanian	On some groups with the zero presentation ranks
Mr. Shripad M. Garge	On conjugacy of Weyl groups
Professor Narain Datt Gupta	A group with the slowest sub-exponential growth
Professor C.K. Gupta	Recent results on automorphism groups
Professor Alessandra Iozzi	Trees and hyperbolic spaces
Dr. D.L. Johnson	Word growth of groups
Professor S.S. Khare	Group actions on smooth manifolds with specific fixed point sets
Dr. Linus Kramer	Symmetric spaces, asymptotic cones and buildings
Professor Ravi Kulkarni	Idea of a branched covering
Mr. Roman Mikhailov	On localization in groups and group varieties
Professor Shahar Mozes	Tree automorphisms, products of trees and lattices
Professor Himadri Mukherjee	On the isomorphism problem and related problems
Professor I.B.S. Passi	Centralizers and homology
Dr. Laszlo Pyber	Boundedly generated groups
Professor M.S. Raghunathan	The word metric on lattices in higher rank semi-simple Lie groups
Dr. A. Raghuram	Depths and conductors of representations of p-adic groups
Dr. Bertrand Remy	Non-linearities of Kac-Moody groups
Professor Mark Ronan	Constructing buildings and twin buildings

Dr. Parameswaran Sankaran	On automorphisms and quasi-isometries of finitely generated groups
Professor Marcus du Sautoy	Counting subgroups in nilpotent groups and points on elliptic curves
Professor Dan Segal	Subgroups of finite index in profinite groups
Professor J.B. Srivastava	Invariants of certain classical and reductive groups and their applications in computer vision
Dr. Geetha Venkataraman	Finite group enumerations: a brief introduction
Professor D.N. Verma	A natural representation model for the symmetric groups, and the ensuring potential for a new unified representation theory for Coxeter Groups
Dr. Christopher Voll	Zeta functions of groups and enumeration in Bruhat-Tits buildings of $Sl_n(Q_p)$
Professor John S. Wilson	Aspects of growth of groups

Conference Participants

Professor Peter Abramenko
Dr. Meenaxi Bhattacharjee
Professor Roger M. Bryant
Professor Marc Burger
Professor Alan Camina
Dr. Indira Chatterji
Professor B.P. Chetiya
Dr. Ahmad Erfanian
Professor C.K. Gupta
Professor Narain Datt Gupta
Professor Alessandra Iozzi
Dr. D.L. Johnson
Professor S.S. Khare
Dr. Linus Kramer
Professor Ravi Kulkarni
Professor Ramji Lal
Professor Shahar Mozes
Professor H.K. Mukherjee
Professor C. Musili
Dr. A. Naolekar
Professor I.B.S. Passi
Professor Dipendra Prasad
Dr. Laszlo Pyber
Dr. K.N. Raghavan
Professor M.S. Raghunathan
Dr. A. Raghuram
Dr. Bertrand Remy
Professor Mark Ronan
Dr. Paraeswaran Sankaran
Dr. Bhaba Kumar Sarma
Professor N.S.N. Sastry
Professor Marcus du Sautoy
Professor Dan Segal
Professor B.L. Sharma
Dr. Ravindra Prasad Shukla
Professor J.B. Srivastava
Dr. B. Sury
Dr. Maneesh Thakur
Professor D.N. Verma
Professor Christopher Voll
Professor John S. Wilson

Workshop Participants

Md. S K Safique Ahmad
Ms. Sasmita Barik
Mr. Dhruba Jyoti Bora
Ms. Ardeline Mary Buhphang
Mr. Anandaram Burhagohain
Mr. Moon Kumar Chetry
Ms. Jonali Choudhury
Mr. Biren Das
Dr. Partha Pratim Dey
Dr. Sanjay Kumar Dutta
Dr. Sanghita Dutta
Mr. Shripad M Garge
Mr. Sahzamal Hoque
Mr. Rafiqul Islam
Mr. Rupesh Kumar Jha
Mr. Bidyut Kalita
Mr. Roman Mikhailov
Mr. Milan Nath
Mr. Tulasi P Nepal
Mr. Binod Kumar Sahoo
Mr. Siddhartha Sarka
Mr. Ritumoni Sarma
Mr. Brajesh Kumar Sharma
Mr. Angom Tiken Singh
Mr. Anupam Kumar Singh
Mr. Ningthoujam Jiban Singh
Mr. Thokchom Vikramjit
Ms. Hiramala Talukdar
Ms. Madhusmita Tripathy
Dr. Geetha Venkataraman
Dr. Christopher Voll
Mr. Akhilesh Chandra Yadav
Mr. Satya Narayan Yadav

Contents

Proc. Inst. Wrk.shps. Int. Conf. – Essays in Geometric Group Theory No. 9, 2009, pp. 1–25.

Linear Representations and Arithmeticity of Lattices in Products of Trees

Marc Burger[1], Shahar Mozes[2] and Robert J. Zimmer[3]

[1]Fim, Eth, 8092 Zürich, Switzerland
[2]Institute of Mathematics, Hebrew University, Jerusalem 91904, Israel
[3]Department of Mathematics, University of Chicago, Chicago, Illinois, USA
e-mail: burger@math.ethz.ch; mozes@math.huji.ac.il; rzimmer@uchicago.edu

1.1. Prolegomenon

In this paper we continue our study of lattices in the automorphisms groups of products of trees initiated in [BM97], [BM00a], [BM00b], [Moz98] (see also [Gla03], [BG02], [Rat04]). We concentrate here on the interplay between the linear representation theory and the structure of these lattices. Before turning to the main results of this paper it may be worthwhile to put certain concepts and results from [BM00a], [BM00b] in perspective, and explain the motivation for our approach.

A lattice Γ in a locally compact group G is a discrete subgroup such that the quotient space G/Γ carries a finite G-invariant measure. If in addition G/Γ is compact the lattice is called cocompact or uniform.

In the framework of this paper the following classical setting will serve as a motivating example. Let $\mathbb{Q}_p$ denote the field of p-adic numbers and given two primes p, q define the locally compact group:

$$G_{p,q} = \mathrm{PSL}(2, \mathbb{Q}_p) \times \mathrm{PSL}(2, \mathbb{Q}_q).$$

Lattices in $G_{p,q}$ fall into two classes: The reducible ones and the irreducible ones. A lattice $\Gamma < G_{p,q}$ is called reducible if it has a subgroup of finite index which is a direct product $\Gamma_p \times \Gamma_q$ where $\Gamma_p < \mathrm{PSL}(2, \mathbb{Q}_p)$ and $\Gamma_q < \mathrm{PSL}(2, \mathbb{Q}_q)$ are lattices. A lattice is called irreducible if it is not reducible. Thus, for a reducible lattice, each of the projections is a closed, in fact discrete, subgroup of the corresponding factor. It is a remarkable fact that for an irreducible lattice $\Gamma < G_{p,q}$ both projections are dense in the corresponding factor. We recall at this point that the notions of reducibility and irreducibility pertain to the theory of lattices in products of semi-simple groups over local fields where analogous density properties hold for irreducible lattices (cf. [Mar91] II Thm. (6.7)). It should be noted that this density result makes essential use of the algebro-geometric structure of the ambient group and in particular of the Borel density theorem. Irreducible lattices in $G_{p,q}$ and more generally in a product G of semisimple groups over local fields of total rank at least 2 enjoy remarkable algebraic and geometric properties as the following results due to G.A. Margulis show:

- *Arithmeticity:* Any irreducible lattice in G is arithmetic. [Mar91, IX Thm (A)].
- *Superrigidity:* Any unbounded irreducible linear representation of an irreducible lattice comes from a rational representation of the ambient group G. [Mar91, VII Thm (5.6)].
- *Normal subgroup Theorem:* Any non-trivial normal subgroup of an irreducible lattice in G is either finite and central or of finite index. [Mar91, IV Thm (4.9)].

An important geometric object associated with $\mathrm{PGL}(2, \mathbb{Q}_p)$ is its Bruhat–Tits building $\mathcal{T}_{p+1}$ which is a regular tree of degree $p + 1$ on which $\mathrm{PGL}(2, \mathbb{Q}_p)$ acts by automorphisms. In fact $\mathrm{PGL}(2, \mathbb{Q}_p)$ acts transitively on the set of vertices of $\mathcal{T}_{p+1}$ and any pair of adjacent vertices is a complete set of representatives for the $\mathrm{PSL}(2, \mathbb{Q}_p)$-orbits. The group $\mathrm{Aut}\mathcal{T}_{p+1}$, with the topology of pointwise convergence on the set of vertices, is a locally compact group containing $\mathrm{PGL}(2, \mathbb{Q}_p)$, and hence $\mathrm{PSL}(2, \mathbb{Q}_p)$, as a closed (cocompact) subgroup. In particular a cocompact lattice in $\mathrm{PGL}(2, \mathbb{Q}_p)$, or $\mathrm{PSL}(2, \mathbb{Q}_p)$, is also one in $\mathrm{Aut}\mathcal{T}_{p+1}$; this allows applying combinatorial and finite group theory techniques, such as the theory of graphs of groups, to the study of the structure of lattices in $\mathrm{PGL}(2, \mathbb{Q}_p)$. As an example we mention the result that a cocompact lattice in $\mathrm{Aut}\mathcal{T}_{p+1}$ is virtually free on finitely many generators (see for example [Ser03]). There are however certain more subtle finiteness properties shared by cocompact lattices in $\mathrm{PGL}(2, \mathbb{Q}_p)$ which fail in general for cocompact lattices in $\mathrm{Aut}\mathcal{T}_{p+1}$. For example one has a positive lower bound on the covolume of lattices in $\mathrm{PGL}(2, \mathbb{Q}_p)$ whereas there are lattices of arbitrarily small covolume in $\mathrm{Aut}\mathcal{T}_{p+1}$. The similarities and differences between the theory of lattices in $\mathrm{Aut}\mathcal{T}_{p+1}$ and the theory of lattices in Lie groups, and in $\mathrm{PGL}(2, \mathbb{Q}_p)$ in particular, are part of an extensive study and we refer to [BL01] for many results and references. Returning to the higher rank situation and in the same geometric vein, we observe that a cocompact lattice Γ in $\mathrm{PGL}(2, \mathbb{Q}_p) \times \mathrm{PGL}(2, \mathbb{Q}_q)$, or in $G_{p,q}$ for that matter, is also a cocompact lattice in $\mathrm{Aut}\mathcal{T}_{p+1} \times \mathrm{Aut}\mathcal{T}_{q+1}$. If Γ is moreover torsion free then it acts freely on the 2-dimensional square complex $\mathcal{T}_{p+1} \times \mathcal{T}_{q+1}$ and the quotient $\Gamma\backslash(\mathcal{T}_{p+1} \times \mathcal{T}_{q+1})$ is a finite square complex inheriting the particular link structure of its universal covering, namely that the link of each vertex is a complete bipartite graph. This puts the study of such lattices in a rich geometric and combinatorial context. In fact, it was observed by D. Wise ([Wis95, Thm 1.5]) that the universal covering space of a 2-dimensional square complex is a product of trees exactly when the link at each vertex is a complete bipartite graph. As for the similarities and differences between the theory of cocompact lattices in $\mathrm{Aut}\mathcal{T}_{p+1} \times \mathrm{Aut}\mathcal{T}_{q+1}$ and $G_{p,q}$, or $\mathrm{PGL}(2, \mathbb{Q}_p) \times \mathrm{PGL}(2, \mathbb{Q}_q)$, the first major difficulty seems to be the proper generalization of the notion of irreducible lattice. This issue will be discussed below after example 1.1.1. Let us indicate one striking phenomenon: while cocompact lattices in $\mathrm{PSL}(2, \mathbb{Q}_p)$, $\mathrm{Aut}\mathcal{T}_{p+1}$ and $\mathrm{PGL}(2, \mathbb{Q}_p) \times \mathrm{PGL}(2, \mathbb{Q}_q)$ are linear groups this need not be so for cocompact lattices in $\mathrm{Aut}\mathcal{T}_{p+1} \times \mathrm{Aut}\mathcal{T}_{q+1}$ and as a consequence of the result proven in this paper we will obtain an algorithmic criterion of non-linearity (see cor. 1.1.20) using the information contained in the finite quotient square complex.

Example 1.1.1. An interesting example of an irreducible lattice in $\mathrm{PGL}(2, \mathbb{Q}_p) \times \mathrm{PGL}(2, \mathbb{Q}_q)$ may be explicitly described as follows. Let p, q be two fixed distinct primes both congruent to 1 modulo 4. Let $\mathbb{H}_{\mathbb{Z}}$ denote the ring of integer quaternions. Fix $\epsilon_p \in \mathbb{Q}_p$ and $\epsilon_q \in \mathbb{Q}_q$, square roots of -1. Define a map $\varphi : \mathbb{H}_{\mathbb{Z}} \backslash \{0\} \to \mathrm{PGL}(2, \mathbb{Q}_p) \times \mathrm{PGL}(2, \mathbb{Q}_q)$ where for $x = x_0 + x_1 i + x_2 j + x_3 k$

$$\varphi(x) = \left(\begin{pmatrix} x_0 + x_1\epsilon_p & x_2 + x_3\epsilon_p \\ -x_2 + x_3\epsilon_p & x_0 - x_1\epsilon_p \end{pmatrix}, \quad \begin{pmatrix} x_0 + x_1\epsilon_q & x_2 + x_3\epsilon_q \\ -x_2 + x_3\epsilon_q & x_0 - x_1\epsilon_q \end{pmatrix} \right)$$

Let $\mathbb{H}_{\mathbb{Z}}(p, q) = \{x \in \mathbb{H}_{\mathbb{Z}} : |x|^2 = x\overline{x} = p^k q^l \text{ for some } k, l \in \mathbb{Z}^+\}$. Its image $\varphi(\mathbb{H}_{\mathbb{Z}}(p, q)) < \mathrm{PGL}(2, \mathbb{Q}_p) \times \mathrm{PGL}(2, \mathbb{Q}_q)$ is an irreducible arithmetic lattice. To avoid torsion we shall be interested in a certain congruence sublattice of it, namely the image of those quaternions in $\mathbb{H}_{\mathbb{Z}}(p, q)$ which are congruent to 1 modulo 2 (i.e., for which x_0 is odd and x_1, x_2 and x_3 are even). Let us denote this lattice by $\Gamma < \mathrm{PGL}(2, \mathbb{Q}_p) \times \mathrm{PGL}(2, \mathbb{Q}_q)$. Using Jacobi's theorem giving the number of ways of representing a number as sum of four squares, one can show that Γ acts freely transitively on the set of vertices of the square complex $\mathcal{T}_{p+1} \times \mathcal{T}_{q+1}$. The resulting quotient $\mathcal{X}_{p,q} = \Gamma \backslash (\mathcal{T}_{p+1} \times \mathcal{T}_{q+1})$ which is a square complex on one vertex admits an explicit description to which we now turn. Let us define the following two sets

$$L_p = \{x \in \mathbb{H}_{\mathbb{Z}} : |x|^2 = p, \quad x \equiv 1 (\mathrm{mod}\ 2) \quad x_0 > 0\} \tag{1.1.1}$$

$$L_q = \{x \in \mathbb{H}_{\mathbb{Z}} : |x|^2 = q, \quad x \equiv 1 (\mathrm{mod}\ 2) \quad x_0 > 0\} \tag{1.1.2}$$

Observe that L_p consists of $p + 1$, and L_q of $q + 1$, elements and if $x \in L_p$ (resp. L_q) then $\overline{x} \in L_p$ (resp. L_q). The 0-skeleton $\mathcal{X}_{p,q}^{(0)}$ consists of a single vertex v_0. The 1-skeleton of $\mathcal{X}_{p,q}$ consists of $\frac{p+1}{2} + \frac{q+1}{2}$ loops based at v_0. We shall partition these loops into two sets E_h and E_v of sizes $\frac{p+1}{2}$ and $\frac{q+1}{2}$ respectively where the elements of E_h will be thought of as horizontal loops and the elements of E_v as vertical loops. We shall label each oriented loop of E_h by an element of L_p where the two orientations of a geometric loop are labeled by conjugate quaternions. Similarly we shall label the oriented loops of E_v by the elements of L_q. Abusing notation we shall refer also to the elements of L_p, L_q as (oriented) loops. To describe the 2-skeleton $\mathcal{X}_{p,q}^{(2)}$, consider all the quadruples $(a, b, \overline{a'}, \overline{b'})$ where $a, a' \in L_p$, $b, b' \in L_q$ and

$$ab = \pm b'a'. \tag{1.1.3}$$

These quadruples come in equivalence classes:

$$\{(a, b, \overline{a'}, \overline{b'}), (\overline{a}, b', a', \overline{b}), (a', \overline{b}, \overline{a}, b'), (\overline{a'}, \overline{b'}, a, b)\}$$

Where the second quadruple is obtained from equation (1.1.3) by multiplying by $\overline{a}$ on the left and $\overline{a'}$ on the right. Similarly the third is obtained by multiplying by $\overline{b'}$ on the left and $\overline{b}$ on the right and finally the last one is obtained by applying quaternionic conjugation to equation (1.1.3).

For each equivalence class we will have a geometric square in $\mathcal{X}_{p,q}^{(2)}$ attached to the 1-skeleton of $\mathcal{X}_{p,q}$ so that when we read the labels of the oriented loops along its boundary in one of the four possible ways starting with an horizontal one we see an element of the corresponding equivalence class. The following arithmetic fact (cf. [Dic22]) is crucial. For each $a \in L_p$ and $b \in L_q$ there are unique $a' \in L_p$ and $b' \in L_q$ so that

$$ab = \pm b'a'.$$

Hence there are $\binom{p+1}{2} \cdot \binom{q+1}{2}$ geometric squares and the link of the vertex v_0 is a complete bipartite graph. Thus, indeed, the universal covering space $\widetilde{\mathcal{X}}_{p,q}$ is a product of a $(p+1)$-regular tree with a $(q+1)$-regular tree. One can then identify $\widetilde{\mathcal{X}}_{p,q}$ with the Bruhat–Tits building $\mathcal{T}_{p+1} \times \mathcal{T}_{q+1}$ associated with $\mathrm{PGL}(2, \mathbb{Q}_p) \times \mathrm{PGL}(2, \mathbb{Q}_q)$ so that $\pi_1(\mathcal{X}_{p,q})$ is identified with Γ.

Let T_1, T_2 be regular trees of finite degrees at least three. We turn to the study of cocompact lattices in a product $\mathrm{Aut}T_1 \times \mathrm{Aut}T_2$. As above we shall call a lattice $\Gamma < \mathrm{Aut}T_1 \times \mathrm{Aut}T_2$ reducible if it has a subgroup of finite index which is a product of lattices in the factors, and irreducible otherwise. In particular any reducible cocompact lattice is virtually a product of two free groups. Let pr_i denote the projection on $\mathrm{Aut}T_i$. Then it is straightforward to verify that for a lattice $\Gamma < \mathrm{Aut}T_1 \times \mathrm{Aut}T_2$, the following are equivalent:

(1) Γ is reducible.
(2) Both subgroups $\mathrm{pr}_1(\Gamma)$ and $\mathrm{pr}_2(\Gamma)$ are closed (hence discrete) in $\mathrm{Aut}T_1$ and $\mathrm{Aut}T_2$.

Given a cocompact lattice Γ we will thus introduce G_i, the closure of $\mathrm{pr}_i(\Gamma)$ in $\mathrm{Aut}T_i$, and for a vertex x_i of T_i let $G_i(x_i)$ denote the stabilizer in G_i of x_i. It is a compact open subgroup of G_i. While $\mathrm{Aut}T_i/G_i$ is compact, the following finiteness result shows that from another point of view, G_i is "far away" from $\mathrm{Aut}T_i$, namely:

Proposition 1.1.2. *Assume that $\Gamma < \mathrm{Aut}T_1 \times \mathrm{Aut}T_2$ is a cocompact lattice. Then the compact group $G_i(x_i)$ is topologically finitely generated.*

Proof. Since $\mathrm{pr}_1(\Gamma)$ is dense in G_1 by definition and $G_1(x_1)$ is open, the subgroup $\mathrm{pr}_1(\Gamma) \cap G_1(x_1)$ is dense in $G_1(x_1)$. On the other hand, we have $\mathrm{pr}_1(\Gamma) \cap G_1(x_1) = \mathrm{pr}_1(\Gamma \cap (G_1(x_1) \times G_2))$. We now proceed to show that $\Gamma \cap (G_1(x_1) \times G_2)$ is finitely generated which will conclude the proof. Let $V := G_1(x_1) \times G_2$ and $G = G_1 \times G_2$. Then V is an open subgroup of G and we claim that $V \cap \Gamma$ is a cocompact lattice in V. Indeed every V-orbit in G/Γ being open is also closed and hence compact. In particular, considering the V-orbit of $e\Gamma$ we obtain a continuous injective map

$$\begin{aligned} V/(V \cap \Gamma) &\to G/\Gamma \\ g(V \cap \Gamma) &\mapsto g\Gamma \end{aligned}$$

which is a homeomorphism on its image, since the image $V\Gamma$ is closed (in fact compact) in G/Γ. Thus $V/(V \cap \Gamma)$ is compact which shows the claim. Finally, $\mathrm{Aut}T_2$ is compactly generated and $\mathrm{Aut}T_2/G_2$ is compact, hence G_2 is compactly generated and hence so is $V := G_1(x_1) \times G_2$. Since $\Gamma \cap V$ is a cocompact lattice in V we deduce that $\Gamma \cap (G_1(x) \times G_2)$ is finitely generated. □

If now T is a regular tree and x a vertex, we have the homomorphism

$$\tau : \mathrm{Aut}T(x) \to (\mathbb{Z}/2\mathbb{Z})^{\mathbb{N}}$$

which associate to each element $g \in \mathrm{Aut}T(x)$ the sequence of signs of the permutations induced by g on all the spheres with center x. If the degree of T is at least 3 it is clear that an arbitrary sequence of signs can specified and obtained by an element $g \in \mathrm{Aut}T(x)$. Hence the homomorphism τ is surjective and it follows in particular that $\mathrm{Aut}T(x)$ is not topologically finitely generated. In particular, cocompact lattices in $\mathrm{Aut}T_1 \times \mathrm{Aut}T_2$ never have dense projections. It would be of interest to understand what happens in the case of non-uniform lattices.

The above discussion leads naturally to the following:

Basic Question. Which groups arise as closures of projections of cocompact lattices in $\mathrm{Aut}T_1 \times \mathrm{Aut}T_2$?

The interest in this question is partly due to a certain number of results showing that there are strong connections between the topological and algebraic properties of the groups $G_i < \mathrm{Aut}T_i$ and the algebraic and geometric properties of Γ; in other words, Γ is as "rigid" as its ambient group $G_1 \times G_2$. In particular, results of this type played a role in [BM00a], [BM00b]. More recently, such connections have been established in a remarkable degree of generality. Monod [Mon05], [Mon06] and Monod–Shalom [MS02] have proved superrigidity results for the actions on CAT(0)-spaces of cocompact lattices $\Gamma < G_1 \times G_2$ in a product of locally compact groups. These results in particular imply the following theorems which we will use in the sequel. In Theorems 1.1.3, 1.1.4, 1.1.5, the groups G_1, G_2 are assumed to be locally compact and compactly generated.

Theorem 1.1.3 ([Mon06, Cor. 4]). *Let $\Gamma < G_1 \times G_2$ be a cocompact lattice with dense projections. Let $\mathbb{H}$ be a connected, k-simple, adjoint k-group where k is a local field and let $\pi : \Gamma \to \mathbb{H}(k)$ be a homomorphism with unbounded Zariski dense image. Then π extends continuously to $G_1 \times G_2$ factoring via one of the factors G_i.*

In certain situations this result will imply that the linear representation theory of Γ over any field is controlled by the continuous representation theory of $G_1 \times G_2$ over local fields.

The next result shows that the various decompositions of Γ as non-trivial amalgams are completely described by the decompositions of G_i, $i = 1, 2$ as non-trivial amalgams. See also [Sha00, Theorem 0.7].

Theorem 1.1.4 ([MS02, Thm. 1.5]). *Let $\Gamma < G = G_1 \times G_2$ be a lattice with dense projections. Let Γ act non-elementarily on a countable tree T. Then there exists an invariant subtree on which the Γ-action extends continuously to a G-action factoring via one of the G_i's.*

Finally the following result of Bader and Shalom (generalizing Margulis' Normal Subgroup Theorem for the case of products) shows that normal subgroups of Γ are in a sense controlled by the closed normal subgroups of $G_1 \times G_2$.

Theorem 1.1.5 ([BS03, Thm. 1.1]). *Let $\Gamma < G = G_1 \times G_2$ be a cocompact lattice with dense projections. Assume that $\mathrm{Hom}_c(G, \mathbb{R}) = (0)$. Let $N \lhd \Gamma$ be a normal subgroup. The group Γ/N is finite if and only if the groups $G_i/\overline{\mathrm{pr}_i(N)}$ are compact.*

Remark 1.1.6. The idea of extending the scope of the rigidity phenomena of lattices in higher rank Lie groups to the general locally compact setting by considering lattices in general products goes at least back to [BnK70] and is present in the work of G. Margulis in the seventies. When G_1 and G_2 are Lie groups over local fields, Theorems 1.1.3, 1.1.4 and 1.1.5 are special cases of results of G. Margulis proven respectively in [Mar74], [Mar81] and [Mar79]. Superrigidity theorems of varying degrees of generality can be found in the book [Mar91] (and references therein), [Bur95], [Mar94], [BM96], [Gao97], [Sha00], [MS02], [Bon], [Mon06] and [GKM08].

If Γ is a cocompact lattice in $\mathrm{Aut}T_1 \times \mathrm{Aut}T_2$ and $G_i = \overline{\mathrm{pr}_i(\Gamma)}$ are, as before, the closures of projections, assuming that Γ is torsion free, we consider the finite square complex $X := \Gamma\backslash(T_1 \times T_2)$. The finite permutation group induced by $G_i(x_i)$ on the set of vertices adjacent to x_i can be explicitly calculated from the combinatorics of X. (We will see an illustration of this later in the paper in the case where X has a single vertex.) It turns out that when Γ is irreducible and this finite permutation group is sufficiently transitive, one obtains rather strong information on the global structure of the non-discrete group G_i. With this in mind we introduced in [BM00a] various classes of closed subgroups defined via "local conditions" and showed how these local conditions imply certain global structure results. We now recall some salient features of this theory.

Let $T = (V, E)$ be a locally finite tree, V its set of vertices and E its set of edges. Given a subgroup $H < \mathrm{Aut}T$ and a vertex $x \in V$, its stabilizer $H(x)$ acts as a finite permutation group on the set $E(x) \subset E$ of edges whose origin is the vertex x, and we say that H is locally quasiprimitive (respectively, locally primitive, locally 2-transitive, etc.) if this finite permutation group is quasiprimitive (respectively primitive, 2-transitive, etc.) for every vertex x. Recall that a finite permutation group $F < \mathrm{Sym}\Omega$ acting on a finite set Ω is called *quasiprimitive* if every non-trivial normal subgroup of F act transitively on Ω. It is called *primitive* if any equivariant factor of Ω is trivial (i.e. either a point or Ω). The action is called 2-transitive if F acts transitively on $\Omega^2\backslash\Delta_\Omega$ where Δ_Ω is the diagonal. Quasiprimitive and primitive groups have a rich structure theory, exemplified by results such as the O'Nan–Scott theorem (cf., [DM96], [Pra97]). Following the classification of the finite simple groups one has also a classification of 2-transitive groups (cf. the survey paper by P. Cameron [Cam95]).

Given a totally disconnected group H let

$$H^{(\infty)} = \cap_{L<H} L$$

where the intersection is taken over all open finite index subgroups. Let

$$\mathrm{QZ}(H) = \{h \in H : Z_H(h) \text{ is open}\}$$

be the quasi-center of H. Both are topologically characteristic subgroups of H. The subgroup $H^{(\infty)}$ is closed, and any normal discrete subgroup of H is contained in

QZ(H). To motivate these definitions observe that when $H = \mathbb{G}(\mathbb{Q}_p)$, where $\mathbb{G}$ is a connected semisimple $\mathbb{Q}_p$-group, then $H^{(\infty)}$ coincides with the subgroup $\mathbb{G}^+$ generated by all one parameter unipotent subgroups of $\mathbb{G}(\mathbb{Q}_p)$, and QZ(H) is the kernel of the adjoint representation of the Lie group $\mathbb{G}(\mathbb{Q}_p)$. We recall next a few basic results established in [BM00a] concerning the structure of these subgroups.

Theorem 1.1.7 ([BM00a] Prop. 1.2.1). *Let T be a locally finite tree. Let $H < \mathrm{Aut}T$ be a closed non-discrete locally quasiprimitive group. Then*

(1) *$H/H^{(\infty)}$ is compact.*
(2) *$\mathrm{QZ}(H)$ is a discrete subgroup of H and $H/\mathrm{QZ}(H)$ is not compact.*
(3) *Any closed normal subgroup of H either contains $H^{(\infty)}$ or is contained in $\mathrm{QZ}(H)$.*

The subgroup QZ(H) acts freely on the set of vertices of T. In particular if QZ(H) is non-trivial then the subgroup (of index at most 2) of QZ(H) consisting of those elements whose translation length is even is a free group on infinitely many generators. In analogy with semisimple Lie groups one would expect the poset of normal subgroups of $H^{(\infty)}$ to have a relatively simple structure; this is indeed the case as shown by theorem 1.1.8, the difficulty being however that local transitivity properties of H are not inherited by $H^{(\infty)}$. A guiding principle in the study of the structure of H is the analogy with the O'Nan-Scot structure theory of finite primitive permutation groups of which theorem 1.1.8 is an analogue. Indeed we have:

Theorem 1.1.8 ([BM00a] Prop. 1.2.1, 1.5.1). *Let $H < \mathrm{Aut}T$ be a closed non-discrete subgroup.*

(1) *If H is locally quasiprimitive, then $H^{(\infty)}/\mathrm{QZ}(H^{(\infty)})$ admits minimal closed normal subgroups. These are finitely many, topologically simple, H-conjugate and their product is dense.*
(2) *If H is locally primitive, $H^{(\infty)}/\mathrm{QZ}(H^{(\infty)})$ is a finite product of topologically simple groups.*
(3) *If H is locally 2-transitive, $H^{(\infty)}/\mathrm{QZ}(H^{(\infty)})$ is topologically simple.*

In all the above cases we have that $\mathrm{QZ}(H^{(\infty)}) = \mathrm{QZ}(H) \cap H^{(\infty)}$ and it acts freely without inversions on T.

We observe that $H^{(\infty)}/\mathrm{QZ}(H^{(\infty)})$ may have an arbitrary large number of simple factors. This is exemplified by Proposition 1.1.15 below and the subsequent discussion. We conclude from Theorem 1.1.8.1 that if H is locally quasiprimitive and $\mathrm{QZ}(H^{(\infty)}) = \{e\}$, then $H^{(\infty)}$ is topologically simple. Indeed, assume $\mathrm{QZ}(H^{(\infty)})$ is trivial; we need to show that there is only one minimal closed normal subgroup in $H^{(\infty)}$. Assume that there were two or more such groups $M_1, M_2, \ldots$; then as all are conjugate and their product is dense they are unbounded and hence M_1 has a non-void limit set in the boundary $T(\infty)$ of the tree. As M_1 is a normal subgroup of $H^{(\infty)}$ and the latter acts cofinitely on the tree, it follows that this limit set must be the whole boundary of T. Since M_1 and M_2 commute this limit set must be pointwise fixed by M_2 which is impossible. We conclude that the unique minimal closed normal subgroup of $H^{(\infty)}$ coincides with $H^{(\infty)}$ showing it is topologically simple. This observation together with Theorem 1.1.5 leads to:

Corollary 1.1.9. *Let $H_i < \mathrm{Aut}T_i$ be closed non-discrete locally quasiprimitive groups and assume $\mathrm{QZ}(H_i) = \{e\}$. Let $\Gamma < \mathrm{Aut}T_1 \times \mathrm{Aut}T_2$ be a cocompact lattice satisfying $H_i^{(\infty)} \subset \overline{\mathrm{pr}_i(\Gamma)} \subset H_i$, $i = 1, 2$. Then any non-trivial normal subgroup of Γ is of finite index.*

Proof. Let $G_i := \overline{\mathrm{pr}_i(\Gamma)}$ and $N \lhd \Gamma$ a normal subgroup. We now verify the hypothesis of Theorem 1.1.5. We will use repeatedly the fact established above that ${H_i}^{(\infty)}$ is topologically simple. We have $\overline{[G_i, G_i]} \supset \overline{[{H_i}^{(\infty)}, {H_i}^{(\infty)}]} = {H_i}^{(\infty)}$, where the last equality follows from the topologically semisimplicity of ${H_i}^{(\infty)}$. Thus, since $H_i/{H_i}^{(\infty)}$ is compact (by theorem 1.1.7 1.) we deduce that $G_i/\overline{[G_i, G_i]}$ is compact as well and hence $\mathrm{Hom}_c(G_i, \mathbb{R}) = (0)$. The subgroup $\overline{\mathrm{pr}_i(N)}$ is normalized by ${H_i}^{(\infty)}$ and hence we have either

(1) $\overline{\mathrm{pr}_i(N)} \cap {H_i}^{(\infty)} = (e)$ or
(2) $\overline{\mathrm{pr}_i(N)} \supset {H_i}^{(\infty)}$.

Assume for instance that (1) occurs for $i = 1$. Then $\overline{\mathrm{pr}_1(N)}$ commutes with ${H_1}^{(\infty)}$ but since ${H_1}^{(\infty)} \backslash T_1$ is finite this implies that $\overline{\mathrm{pr}_1(N)} = (e)$. This implies that $\mathrm{pr}_2(N)$ is a discrete subgroup of $\mathrm{Aut}T_2$, normalized by ${H_2}^{(\infty)}$ and thus $\mathrm{pr}_2(N) \cap {H_2}^{(\infty)} \subset QZ({H_2}^{(\infty)}) = (e)$. Thus again, $\mathrm{pr}_2(N) = (e)$ which implies $N = (e)$. Thus if $N \neq (e)$ then $\overline{\mathrm{pr}_i(N)} \supset {H_i}^{(\infty)}$ for $i = 1, 2$ and hence $G_i/\overline{\mathrm{pr}_i(N)}$ is compact which together with Theorem 1.1.5 implies that Γ/N is finite. □

Remark 1.1.10. If Γ satisfies the conditions of Corollary 1.1.9 then so does also any finite index subgroup of Γ. See the beginning of §1.3.

The quasi-center plays, in several ways, an important role for the structure of Γ. In [BM00b], Theorem 1.1.7 was applied to obtain the following criterion for non-residually finiteness of Γ. Let $\Lambda_1 = \Gamma \cap (\mathrm{Aut}T_1 \times e)$, $\Lambda_2 = \Gamma \cap (e \times \mathrm{Aut}T_2)$. We have:

Proposition 1.1.11 ([BM00b] Prop. 2.1, 2.2). *Let $H_i < \mathrm{Aut}T_i$ be closed, non-discrete, locally quasiprimitive groups and $\Gamma < \mathrm{Aut}T_1 \times \mathrm{Aut}T_2$ be a lattice with $H_i^{(\infty)} \subset \overline{\mathrm{pr}_i(\Gamma)} \subset H_i$. Then*

(1) *If $\Lambda_1 \cdot \Lambda_2 \neq \{e\}$ then Γ is not residually finite.*
(2) *If $H_i = \overline{\mathrm{pr}_i(\Gamma)}$ then Λ_i is a normal subgroup of $\mathrm{QZ}(H_i)$ and the quotients $\mathrm{QZ}(H_i)/\Lambda_i$ are locally finite groups.*

In the setting of Proposition 1.1.11.2, we have that Λ_i is non-trivial if and only if the quasi center $\mathrm{QZ}(H_i)$ is non-trivial. Theorems 1.1.5, 1.1.7 and Proposition 1.1.11.2 imply (see 1.3.2):

Theorem 1.1.12. *Let $\Gamma < \mathrm{Aut}T_1 \times \mathrm{Aut}T_2$ be a cocompact lattice such that $H_i = \overline{\mathrm{pr}_i(\Gamma)}$ are non-discrete, locally quasiprimitive. Let $N \lhd \Gamma$ be a normal subgroup. Then either*

1. *$N \subset \mathrm{QZ}(H_1) \times \mathrm{QZ}(H_2)$, or*
2. *N is of finite index in Γ.*

Theorem 1.1.11 together with a predecessor (see [BM00b] §4) of the Normal Subgroup Theorem of Bader–Shalom (Theorem 1.1.5) enabled us to construct a new class of finitely presented simple groups.

Theorem 1.1.13 ([BM97], [BM00b]). *For all sufficiently large $n, m \in \mathbb{N}$ there is a torsion-free cocompact lattice in* $\mathrm{Aut}T_{2n} \times \mathrm{Aut}T_{2m}$ *which is a simple group.*

Remark 1.1.14. In [Rat04] Diego Rattaggi has constructed many interesting lattices $\Gamma < \mathrm{Aut}T_1 \times \mathrm{Aut}T_2$. In particular he has constructed a (4, 12)-complex ([Rat04, Example 2.26]) whose fundamental group is not residually finite (using Proposition 1.1.11). Using this example as well as his construction of an (A_6, A_6)-square complex described in Example 1.1.19 and following the same strategy as in [BM00b] he has established the existence of simple groups as above for all $n \geq 9$ and $m \geq 13$, see [Rat04, Prop. 2.29 (1)].

An important ingredient in the proof of Theorem 1.1.13 consists of the identification of the closures of the projections of lattices constructed via geometric methods. This brings us back to the basic question above of which groups arise as closures of projections of uniform lattices in $\mathrm{Aut}T_1 \times \mathrm{Aut}T_2$.

The finitely generated simple groups of Theorem 1.1.13 are of course non-linear.

In the rest of the paper we will be interested in studying lattices which admit infinite linear images. We will show that such lattices are essentially extensions of arithmetic lattices in semisimple Lie groups over appropriate local fields.

In view of the superrigidity theorems mentioned above, locally quasiprimitive groups which admit a p-adic analytic structure will play a central role in the discussion of arithmeticity. We have:

Proposition 1.1.15. *Let $H < \mathrm{Aut}T$ be a closed non-discrete locally quasiprimitive group. Assume that it admits a $\mathbb{Q}_p$-analytic structure. Let $\mathfrak{H}$ denote the Lie algebra of H, let $\mathbb{G} = \mathrm{Aut}(\mathfrak{H} \otimes \overline{\mathbb{Q}_p})$ a linear algebraic group defined over $\mathbb{Q}_p$ and let $\mathrm{Ad} : H \to \mathbb{G}(\mathbb{Q}_p)$ be the adjoint representation. Then*

(1) *$\mathbb{G}$ is adjoint and semisimple.*
(2) $\ker \mathrm{Ad} = \mathrm{QZ}(H)$.
(3) $\mathrm{Ad}(H) \supset \mathbb{G}^+$.

As mentioned above $\mathbb{G}^+$ denotes for a $\mathbb{Q}_p$-group $\mathbb{G}$, the subgroup of $\mathbb{G}(\mathbb{Q}_p)$ generated by all the one parameter unipotent $\mathbb{Q}_p$-subgroups.

Thus the group H is an extension by $\mathrm{QZ}(H)$ of the group $\mathrm{Ad}(H)$ which lies between $\mathbb{G}^+$ and $\mathbb{G}(\mathbb{Q}_p)$. Recall that $\mathrm{QZ}(H)$ is either trivial or virtually a free group on infinitely many generators. We remark that even though H, being a subgroup of $\mathrm{Aut}T$, is a "rank one object" the group $\mathbb{G}$ may be of arbitrarily large rank. Consider a connected simple algebraic group $\mathbb{G}$ defined over $\mathbb{Q}_p$ and let Δ be its affine Bruhat–Tits building. The action of $\mathbb{G}$ on Δ induces an action on any graph $\mathcal{G}$ which is equivariantly drawn on Δ and this leads to the extension of $\mathbb{G}$ by $\pi_1(\mathcal{G})$ acting on the universal covering tree $\tilde{\mathcal{G}}$. In particular if one takes $\mathcal{G}$ to be the subgraph of the 1-skeleton of Δ obtained by considering all edges of a given colour, then the resulting extension H, $1 \to \pi_1(\mathcal{G}) \to H \to \mathbb{G}(\mathbb{Q}_p) \to 1$ is a locally primitive group.

Observe that in the setting of Proposition 1.1.15 the homomorphism $\mathrm{Ad} : H \to \mathbb{G}(\mathbb{Q}_p)$ sends every vertex stabilizer $H(x)$, x a vertex of T, isomorphically to a compact subgroup of $\mathbb{G}(\mathbb{Q}_p)$. In particular $H(x)$ is a virtually pro-p-group, i.e., $H(x)$ contains an open pro-p-subgroup of finite index. However in general it is not clear to which extent the homomorphism is implemented by a geometric mapping of the tree into the Bruhat–Tits building associated to $\mathbb{G}(\mathbb{Q}_p)$, and thus we have no real way of comparing the "congruence" structure on $H(x)$ induced by the tree structure of T and the congruence structure induced by that of a maximal compact subgroup of $\mathbb{G}(\mathbb{Q}_p)$. There is however a powerful tool going back to Thompson and Wielandt:

Proposition 1.1.16 ([BM00a], Prop. 2.1.2). *Let $T = (V, E)$ be a locally finite tree. Let $H < \mathrm{Aut}T$ be a closed locally primitive subgroup and when H is non-discrete assume that $H(x)$ is virtually pro-p. Then*

(1) *There is a vertex $y \in V$ such that $H_2(y)$ is pro-p.*
(2) *If H acts transitively on V, then for any two adjacent vertices x, y the subgroup $H_1(x, y) = H_1(x) \cap H_1(y)$ is pro-p.*

Where as in [BM00a] $H_i(x)$ denotes the pointwise stabilizer of the ball of radius i centered at x.

Our main result is:

Theorem 1.1.17. *Let T_1, T_2 be locally finite trees and $\Gamma < \mathrm{Aut}T_1 \times \mathrm{Aut}T_2$ a cocompact lattice. Assume*

(1) $H^{(\infty)} < \overline{\mathrm{pr}_i(\Gamma)} < H_i$, *where $H_i < \mathrm{Aut}T_i$ is a closed non-discrete, locally quasi-primitive subgroup.*
(2) *There is a linear representation $\pi : \Gamma \to \mathrm{GL}(n, \mathbb{C})$ with infinite image.*

Then there are prime numbers p_1, p_2 such that for each $i = 1, 2$ H_i is $\mathbb{Q}_{p_i}$-analytic. Let $\mathbb{G}_i$ be the adjoint semisimple algebraic group defined over $\mathbb{Q}_{p_i}$ given by Proposition 1.1.15 and $\mathrm{Ad}_i : H_i \to \mathbb{G}_i(\mathbb{Q}_{p_i})$ the adjoint representation. Then

(1) $(\mathrm{Ad}_1 \times \mathrm{Ad}_2)(\Gamma)$ *is an arithmetic lattice in* $\mathbb{G}_1(\mathbb{Q}_{p_1}) \times \mathbb{G}_2(\mathbb{Q}_{p_2})$.
(2) $\Lambda_i := \Gamma \cap H_i$ *is of finite index in* $\mathrm{QZ}(H_i) := \ker \mathrm{Ad}_i$.

and Γ fits into the exact sequence

$$1 \to \Lambda_1 \times \Lambda_2 \to \Gamma \to (\mathrm{Ad}_1 \times \mathrm{Ad}_2)(\Gamma) \to 1$$

Using the above result we can now characterize the "classical" situation:

Corollary 1.1.18. *Let T_1, T_2 be locally finite trees. Let $\Gamma < \mathrm{Aut}T_1 \times \mathrm{Aut}T_2$ be a cocompact lattice. Assume*

(1) $H_i^{(\infty)} < \overline{\mathrm{pr}_i(\Gamma)} < H_i$, *where $H_i < \mathrm{Aut}T_i$ is a closed non-discrete, locally primitive subgroup.*
(2) *There is a linear representation $\pi : \Gamma \to \mathrm{GL}(n, \mathbb{C})$ with infinite image.*

Then the following are equivalent:

(1) Γ *is linear over* $\mathbb{C}$.
(2) Γ *is residually finite.*
(3) $\text{rank}_{\mathbb{Q}_{p_i}}(\mathbb{G}_i) = 1$ *for both* $i = 1, 2$.

In this case the geometric realization $|T_i|$ *is isometric to the Bruhat–Tits tree associated to* $\mathbb{G}_i$.

We turn next to discuss some consequences of the above results formulated in geometric terms, that is expressed in terms of finite square complexes $\Gamma\backslash(T_1 \times T_2)$ rather than in terms of lattices $\Gamma < \text{Aut}T_1 \times \text{Aut}T_2$. We will concentrate here on the special case of a square complex with a single vertex. Such square complexes correspond to lattices acting freely transitively on the set of vertices of the product of trees $T_1 \times T_2$. A square complex with one vertex whose universal covering space is a product of two trees is given by the following data (see Example 1.1.1):

– Two finite sets A_1, A_2 each endowed with a fixed point free involution $x \mapsto \overline{x}$.
– A set of "squares" $S \subset A_1 \times A_2 \times A_1 \times A_2$ satisfying the following two conditions:

(1) The elements of S come in equivalence classes:

$$\{(a, b, a', b'), (\overline{a}, \overline{b'}, \overline{a'}, \overline{b}), (\overline{a'}, \overline{b}, \overline{a}, \overline{b'}), (a', b', a, b)\} \subset S$$

(2) *(Link condition)* For each pair $(a, b) \in A_1 \times A_2$ there is a unique pair $(a', b') \in A_1 \times A_2$ such that $(a, b, a', b') \in S$.

As in Example 1.1.1 this data determines a 2-dimensional 1-vertex square complex X with a single vertex denoted x_0, $n_1 = \frac{1}{2}|A_1|$ horizontal geometric loops, $n_2 = \frac{1}{2}|A_2|$ vertical geometric loops and $n_1 \cdot n_2 = \frac{1}{4}|S|$ geometric squares. The complex X is the quotient $\Gamma\backslash(T_1 \times T_2)$ where each T_i is a $2n_i$ regular tree and $\Gamma = \pi_1(X, x_0)$. The complex X provides also a presentation for Γ, namely:

$$\Gamma = \langle A_1 \cup A_2 \mid aba'b',\ (a, b, a', b') \in S,\ a\overline{a}, a \in A_1,\ b\overline{b},\ b \in A_2\rangle.$$

The local transitivity properties of Γ on each factor can be easily read off the complex X by considering the holonomy action of elements corresponding to oriented horizontal loops on vertical edges and vice versa. Fixing a vertex $(x_1, x_2) \in T_1 \times T_2$ the group $\langle A_1\rangle$, which is a free group on $\frac{1}{2}|A_1|$ generators, may be identified with the stabilizer $\text{Stab}_\Gamma(x_2)$. The covering map $T_1 \times T_2 \to X$ induces a labeling of the oriented 1-skeleton of $T_1 \times T_2$ by the elements of $A_1 \cup A_2$. We may identify T_1 with the tree consisting of the connected component of the "horizontal 1-skeleton" of $T_1 \times T_2$ containing the chosen vertex (x_1, x_2). This induces a labeling of the oriented edges of T_1 by the elements of A_1 and similarly we have a labeling of the oriented edges of T_2 by the elements of A_2. Observe that paths without backtracking of length k starting at the vertex $x_2 \in T_2$ correspond to $A_2^{(k)}$ which is the set of irreducible words of length k over A_2. The action of $\langle A_1\rangle$ on $A_2^{(k)}$, that is the permutation representation $\rho_{12}^{(k)} : \langle A_1\rangle \to \text{Sym}A_2^{(k)}$ can be read directly from the complex by observing that for $a \in A_1$ and $w \in A_2^{(k)}$ there exist a unique pair: $a' \in A_1$, $w' \in A_2^{(k)}$ such that $a \cdot w' = w \cdot a'$. We have $\rho_{12}^{(k)}(a)(w) = w'$. Exchanging the roles of the trees and the generating sets we obtain: $\rho_{21}^{(k)} : \langle A_2 \rangle \to \mathrm{Sym} A_1^{(k)}$. Let us define

$$P_2^{(k)} = \mathrm{Im}\, \rho_{12}^{(k)} < \mathrm{Sym} A_2^{(k)}$$
$$P_1^{(k)} = \mathrm{Im}\, \rho_{21}^{(k)} < \mathrm{Sym} A_1^{(k)}$$

We shall omit the superscript (k) when $k = 1$. We will say that X is a (P_1, P_2)-complex when we want to emphasize the local permutation groups $P_1 < \mathrm{Sym} A_1$ and $P_2 < \mathrm{Sym} A_2$.

Example 1.1.19 ([Rat04] Theorem 2.3). The (A_6, A_6)-complex X given by:

$$\{a_1 b_1 \overline{a_1} \overline{b_1}, a_1 b_2 \overline{a_1} \overline{b_3}, a_1 b_3 a_2 \overline{b_2}, a_1 \overline{b_3} \overline{a_3} b_2, a_2 b_1 \overline{a_3} \overline{b_2},$$
$$a_2 b_2 \overline{a_3} \overline{b_3}, a_2 b_3 \overline{a_3} b_1, a_2 \overline{b_3} a_3 b_2, a_2 \overline{b_1} \overline{a_3} \overline{b_1}\}$$

Satisfies the following:

(1) Any non-trivial normal subgroup of $\pi_1(X)$ is of finite index.
(2) Any linear representation of $\pi_1(X)$ in characteristic zero has finite image.
(3) $\mathrm{Out}(\pi_1(X)) \equiv \mathbb{Z}/2\mathbb{Z}$.

We will say that a complex X is irreducible if the lattice $\Gamma = \pi_1(X)$ is irreducible, namely when no finite covering of X is a product of two graphs. It is a fundamental problem whether there is an algorithm for deciding if a given X is irreducible. There is however a sufficient condition based on the Thompson-Wielandt Theorem, see Proposition 1.1.16, which we describe next. Assume that $P_i < \mathrm{Sym} A_i$ is transitive, $i = 1, 2$. Fix an edge e_i in T_i and let $B_r(e_i) \subset T_i$ denote the neighbourhood of radius r around e_i, let Q_i denote the restriction to $B_2(e_i)$ of the subgroup of $P_i^{(3)}$ consisting of elements pointwise fixing $B_1(e_i)$. It follows from the Thompson-Wielandt theorem 1.1.16 that if for some $i = 1, 2$ $Q_i \neq \{e\}$ and is not a p-group then the lattice is irreducible. In fact as a corollary of the arithmeticity theorem a much stronger assertion holds:

Corollary 1.1.20. *Let X be a (P_1, P_2)-complex where $P_i < \mathrm{Sym} A_i$ are primitive permutation groups, $i = 1, 2$, and assume that $Q_1 \neq \{e\}$ and is not a p-group. Then any linear representation of $\pi_1(X)$ over a field of characteristic zero has finite image.*

In applying the criterion for non-residually finiteness (Proposition 1.1.11) one needs to detect whether the projection of Γ on one of the factor, say $\mathrm{Aut} T_2$, has a non-trivial kernel. In other words whether there exists an element $w \in \langle A_1 \rangle$ whose action on T_2 is trivial. Observe that while verifying whether a given element $w \in \langle A_1 \rangle$ acts trivially on T_2 is a (easily) computable question, we do not know whether the existence of such an element is decidable. We also do not know an algorithm for deciding whether the quasi-center of $H_1 = \overline{\mathrm{pr}_1(\Gamma)}$ is trivial or not. There is however, a construction – called fibered product, see [BM00b] §2.2, 2.3 – which starting with a 1-vertex square complex X produces a new 1-vertex square complex $X \boxtimes X$ which in suitable settings has a non-residually finite fundamental group whose closures of projections have non-trivial quasi-centers. Let X be given by the data: A_1, A_2 and $S \subset A_1 \times A_2 \times A_1 \times A_2$ then $X \boxtimes X$ is defined by the data: $A_1 \times A_1$, $A_2 \times A_2$ and $R = \{((a_1, a_2), (b_1, b_2), (a'_1, a'_2), (b'_1, b'_2)) : (a_i, b_i, a'_i, b'_i) \in S,\ i = 1, 2\}$. If X is a (P_1, P_2)-complex then $X \boxtimes X$ is a $(P_1 \times P_1, P_2 \times P_2)$-complex and hence $\pi_1(X \boxtimes X)$ can never have locally primitive projections. However, see [BM00b] §2.2, it is contained as an index 2 subgroup in a group which is locally primitive when P_1 and P_2 are primitive. The natural homomorphism

$$\pi_1(X \boxtimes X) \to \pi_1(X) \times \pi_1(X) \tag{1.1.4}$$

has image of index 1, 2 or 4 and its kernel defines an infinite covering which is of the form $\mathcal{D}_1 \times \mathcal{D}_2$ where each $\mathcal{D}_i$ is a graph whose fundamental group is an infinitely generated free group. Using [BM00b] Prop. 2.1 we showed:

Proposition 1.1.21. *If X is an irreducible (P_1, P_2)-square complex where each P_i is a 2-transitive permutation group with 2-transitive socle*[1] *then $\pi_1(X \boxtimes X)$ is not residually finite; in fact $\pi_1(X \boxtimes X)^{(\infty)} \supset \pi_1(\mathcal{D}_1) \times \pi_1(\mathcal{D}_2)$.*

Examples of 2-transitive permutation groups with 2-transitive socle include S_n, A_n $(n \geq 5)$, PSL$(2, \mathbb{F}_q)$. See [Cam95, Table 5.1(b)] and [DM96, 7.7] for a complete list of such groups. Composing the homomorphism in (1.1.4) with the projection on either factor we obtain a surjective homomorphism, $\pi_1(X \boxtimes X) \to \pi_1(X)$ showing that rigidity fails in a strong way since not even the universal covering spaces of $X \boxtimes X$ and X are isomorphic. However in §1.4 we will show:

Corollary 1.1.22. *Let X be an irreducible (P_1, P_2)-complex and Y a (Q_1, Q_2)-complex. Assume that P_1 and P_2 are primitive permutation groups. Then any surjective homomorphism $\varphi : \pi_1(X) \to \pi_1(Y)$ is induced by an isomorphism $f : X \to Y$. In particular φ is an isomorphism and each Q_i is permutation isomorphic to P_i (up to reindexing).*

1.2. Locally quasiprimitive groups and p-adic structure

The basic objective of this chapter is to show that a locally quasiprimitive group H which admits a continuous representation into GL$(n, \mathbb{Q}_p)$ with unbounded image is p-adic analytic, and to investigate this p-adic structure more closely. First we establish a general fact about continuous representations of totally disconnected groups. This will rest on the following consequence of the Howe-Moore Theorem [HM79]. We have:

Theorem 1.2.1. *Let $\mathbb{G}$ be a $\mathbb{Q}_p$-almost simple group and $O < \mathbb{G}(\mathbb{Q}_p)$ an open unbounded subgroup. Then $O \supset \mathbb{G}^+$*

Proof. We may clearly assume that $\mathbb{G}$ is connected. Then $\delta_O \in \ell^2(\mathbb{G}(\mathbb{Q}_p)/O)$ is a vector which is invariant under the unbounded group O and hence by the Howe-Moore theorem is $\mathbb{G}^+$-invariant. □

[1] The *socle* is the subgroup generated by all minimal normal subgroups.

Remark 1.2.2. To prove Theorem 1.2.1 one needs in fact only the local field analogue of Mautner phenomenon — as proven by C. Moore for semisimple real Lie groups in [Moo66, Theorem 1]. The arguments of Howe–Moore in [HM79, Proposition 5.5] can be used to give a direct proof of the local field analogue of [Moo66, Theorem 1] from which Theorem 1.2.1 follows.

Lemma 1.2.3. *Let H be a locally compact totally disconnected group, $K < H$ an open compact subgroup and $\mathbb{G}$ a $\mathbb{Q}_p$-almost simple group. Let $\pi : H \to \mathbb{G}(\mathbb{Q}_p)$ be a continuous homomorphism with Zariski dense image. Then one of the following holds:*

(1) *$\pi(K)$ is finite.*
(2) *$\pi(K)$ is open and $\pi(H)$ is compact.*
(3) *$\pi(K)$ is open and $\pi(H) \supset \mathbb{G}^+$.*

Proof. Denote by $\mathfrak{G}$ the Lie algebra of $\mathbb{G}$. If the closed subgroup $\pi(K)$ is infinite, its Lie algebra $\mathcal{K} \subset \mathfrak{G}(\mathbb{Q}_p)$ is of positive dimension. As $\pi(H)$ commensurates $\pi(K)$ it follows that $\operatorname{Ad}\pi(H)$ leaves $\mathcal{K}$ invariant. Hence by the Zariski density of $\pi(H)$ it follows that $\mathcal{K} = \mathfrak{G}(\mathbb{Q}_p)$, implying that $\pi(K)$ is open. If $\pi(H)$ is not compact it is open and unbounded and hence by Theorem 1.2.1 we have $\pi(H) \supset \mathbb{G}^+$. □

Before proceeding we need a few more results from [BM00a] concerning locally quasiprimitive groups.

Proposition 1.2.4 (see [BM00a, Prop. 1.2.1]). *Let T be a tree and $H < \operatorname{Aut}T$ a closed non-discrete locally quasiprimitive subgroup. we have:*

(1) *$\mathrm{QZ}(H^{(\infty)})$ is discrete, in fact, $\mathrm{QZ}(H^{(\infty)}) = H^{(\infty)} \cap \mathrm{QZ}(H)$.*
(2) *for any open normal subgroup $N H^{(\infty)}$ we have $N = H^{(\infty)}$.*
(3) *$H^{(\infty)} = \overline{[H^{(\infty)}, H^{(\infty)}]}$.*

Lemma 1.2.3 applied to locally quasiprimitive groups implies the following:

Lemma 1.2.5. *Let $H < \operatorname{Aut}T$ be a closed non-discrete locally quasiprimitive group. Let L be a closed subgroup, $H^{(\infty)} < L < H$, $\mathbb{G}$ an almost $\mathbb{Q}_p$-simple group and $\pi : L \to \mathbb{G}(\mathbb{Q}_p)$ a continuous homomorphism with Zariski dense image. Then either*

(1) *$\pi(L)$ is compact and hence $\pi(H^{(\infty)}) = \{e\}$. or*

(2) (a) $\pi(H^{(\infty)}) = \mathbb{G}^+$,
 (b) $\pi(\mathrm{QZ}(H^{(\infty)})) \subset Z(\mathbb{G}(\mathbb{Q}_p))$,
 (c) *$\pi(K)$ is open for any compact open subgroup $K < H^{(\infty)}$.*

Proof. Assume that $\pi(L)$ is not compact, then $\pi(H^{(\infty)})$ is not finite (Theorem 1.1.7.1) in particular not central and being normal in a Zariski dense subgroup of an almost $\mathbb{Q}_p$-simple group it is itself Zariski dense. Now we apply the trichotomy of Lemma 1.2.3 to π, $H^{(\infty)}$ and K. We have:

(1) If $\pi(K)$ is finite then $\ker\pi \lhd H^{(\infty)}$ is open and hence (by Prop. 1.2.4.2) we have $\ker\pi = H^{(\infty)}$ and (by Theorem 1.1.7.1) hence $\pi(L)$ is compact, contradiction.
(2) If $\pi(K)$ is open and $\pi(H^{(\infty)})$ is compact then $\pi(H^{(\infty)})$ is profinite; by Proposition 1.2.4.2 any profinite quotient of $H^{(\infty)}$ is trivial, hence $\pi(H^{(\infty)}) = \{e\}$, thus $\pi(L)$ is compact, contradiction.

(3) We have that $\pi(K)$ is open and $\pi(H^{(\infty)}) \supset \mathbb{G}^+$. Since $\mathbb{G}/\mathbb{G}^+$ is finite one has $\pi(H^{(\infty)}) = \mathbb{G}^+$. Finally $\pi(\mathrm{QZ}(H^{(\infty)}))$ is a countable subgroup of $\mathbb{G}(\mathbb{Q}_p)$ normalized by $\mathbb{G}^+$ hence contained in $Z(\mathbb{G}(\mathbb{Q}_p))$. □

The preceding lemma gives useful information about homomorphisms of locally quasiprimitive groups into almost simple groups. The next lemma says that in a continuous representation there is always a semisimple part.

Lemma 1.2.6. *Let H be a non-discrete, locally quasiprimitive group. Let $\pi : H^{(\infty)} \to \mathrm{GL}(n, \mathbb{Q}_p)$ be a continuous non-trivial representation and $\mathbb{L}$ be the Zariski closure of $\pi(H^{(\infty)})$. Then*

(1) *$\mathbb{L}$ is connected.*
(2) *$\mathbb{L}/\mathrm{Rad}(\mathbb{L})$ is of positive dimension.*

Proof. Since $H^{(\infty)}$ does not have non-trivial continuous finite images it follows that $\mathbb{L}^\circ = \mathbb{L}$ and $\mathbb{L}$ is connected. Assertion 2 follows from the fact that $\mathbb{L}$ is connected and $H^{(\infty)}$ is topologically perfect (see Prop. 1.2.4.3). □

It follows that if there exists a non-trivial continuous representation $\pi : H^{(\infty)} \to \mathrm{GL}(n, \mathbb{Q}_p)$ then there exists a $\mathbb{Q}_p$-simple, connected, adjoint group $\mathbb{G}$ of positive dimension and a continuous homomorphism $\rho_0 : H^{(\infty)} \to \mathbb{G}(\mathbb{Q}_p)$ with Zariski dense image. Thus we can use Lemma 1.2.5 with $L = H^{(\infty)}$. Observe that $\rho_0(H^{(\infty)})$ cannot be trivial, hence

- $\rho_0(H^{(\infty)}) = \mathbb{G}^+$
- $\mathrm{QZ}(H^{(\infty)}) \subset \ker \rho_0$

We recall a few results from [BM00a] concerning normal subgroups of $H^{(\infty)}$. Let $\mathcal{M}$ be the set of closed normal subgroups M of $H^{(\infty)}$, which are minimal with respect to the property $M \supsetneq \mathrm{QZ}(H^{(\infty)})$. Then (see Theorem 1.1.8.1) the set $\mathcal{M}$ is non-void and contains finitely many elements. These elements are H-conjugate and their product is dense in $H^{(\infty)}$.

In our setting notice that since $\mathrm{QZ}(H^{(\infty)}) \subset \ker \rho_0 \subsetneq H^{(\infty)}$ it follows from Theorem 1.1.8.1 that there exists $M_1 \in \mathcal{M}$, such that $M_1 \not\subset \ker \rho_0$. Let also $h_1 = e, h_2, \dots, h_n \in H$ be such that

$$\mathcal{M} = \{h_i M_1 h_i^{-1} : 1 \leq i \leq n\}$$

and define $\rho_i : H^{(\infty)} \to \mathbb{G}(\mathbb{Q}_p)$ by $\rho_i(h) = \rho_0(h_i^{-1} h h_i)$. Then the following holds:

Lemma 1.2.7. *The homomorphism $\rho : H^{(\infty)} \to (\mathbb{G}^+)^n$ given by $\rho(h) = (\rho_i(h))_{1 \leq i \leq n}$ induces a topological isomorphism*

$$H^{(\infty)}/\mathrm{QZ}(H^{(\infty)}) \to (\mathbb{G}^+)^n.$$

Proof. Since $\mathbb{G}^+$ is simple and $\rho_0(M_1)$ is a non-trivial normal subgroup of $\rho_0(H^{(\infty)}) = \mathbb{G}^+$ we have that $\rho_0(M_1) = \mathbb{G}^+$ and hence $M_1 \ker \rho_0 = H^{(\infty)}$. We claim that $M_2 \cdots M_n \subset \ker \rho_0$. Since $M_1 \cap M_i = \mathrm{QZ}(H^{(\infty)})$ (by minimality), it follows that for $2 \leq i \leq n$, $[M_1, M_i] \subset \mathrm{QZ}(H^{(\infty)})$ and hence $\rho_0(M_1) = \mathbb{G}^+$ commutes with

$\rho_0(M_i)$ which implies $\rho_0(M_i) = e$, $2 \le i \le n$, and the claim is proved. We conclude that $\rho(M_1 \cdots M_n) = (\mathbb{G}^+)^n$. Note also $\ker \rho \supset \mathrm{QZ}(H^{(\infty)})$ and in fact one has equality since otherwise (see [BM00a] Prop. 1.5.1(3)) $\ker \rho$ would contain one of the M_i's. □

The following is one of the main results of this chapter:

Theorem 1.2.8. *Let $H < \mathrm{Aut}T$ be a closed non-discrete locally quasiprimitive subgroup, and let $H^{(\infty)} < L < H$ be a closed subgroup. Assume that there exists a continuous, unbounded representation:*

$$\pi : L \to \mathrm{GL}(m, \mathbb{Q}_p).$$

Then

(1) *H is a $\mathbb{Q}_p$-analytic group and $H^{(\infty)}$ is of finite index in H.*
(2) *Its Lie algebra is a product $\mathrm{Lie}(H) = \mathfrak{G}^n$ where $\mathfrak{G}$ is a $\mathbb{Q}_p$-simple Lie algebra.*
(3) *The adjoint representation $\mathrm{Ad} : H \to \mathrm{Aut}(\mathfrak{G}^n)$ has kernel equal $\mathrm{QZ}(H)$ and induces an isomorphism: $H^{(\infty)}/\,\mathrm{QZ}(H^{(\infty)}) \to (\mathbb{G}^+)^n$ where $\mathbb{G}$ is the connected $\mathbb{Q}_p$-simple adjoint group attached to $\mathfrak{G}$.*

Proof. Applying Lemmas 1.2.5, 1.2.6 we are in the setting of Lemma 1.2.7. As both $H^{(\infty)}$ and $\mathrm{QZ}(H^{(\infty)})$ are normalized by H, H acts by conjugation on $H^{(\infty)}/\,\mathrm{QZ}(H^{(\infty)})$ and we obtain via ρ a continuous action of H on $(\mathbb{G}^+)^n$, which we denote $\epsilon : H \to \mathrm{Aut}((\mathbb{G}^+)^n)$, with $\epsilon(H^{(\infty)}) = \mathrm{Inn}((\mathbb{G}^+)^n)$. The group $H' = \epsilon^{-1}(\mathrm{Inn}((\mathbb{G}^+)^n))$ is a closed subgroup of finite index in H. Then $H^{(\infty)} \cdot \ker \epsilon = H'$, but since $\ker \epsilon$ is a closed normal subgroup which is not cocompact, we deduce (using Prop. 1.1.7.1) that $\ker \epsilon \subset \mathrm{QZ}(H)$, in particular $H^{(\infty)}$ is of countable index in H and hence (by Baire category) of finite index. Assertion (1) is proved. The remaining statements follow by repeated application of Lemma 1.2.7. □

We finish this section with a result which gives information on homomorphisms from a totally disconnected group into a semisimple group over $\mathbb{Q}_p$.

Proposition 1.2.9. *Let H be a locally compact, totally disconnected second countable group, $\mathbb{G}_1$, $\mathbb{G}_2$ connected adjoint semisimple $\mathbb{Q}_p$-groups such that $\mathbb{G}_2$ is $\mathbb{Q}_p$-simple and every $\mathbb{Q}_p$-simple factor of $\mathbb{G}_1$ has positive $\mathbb{Q}_p$-rank. Let*

$$\pi : H \to \mathbb{G}_1(\mathbb{Q}_p) \times \mathbb{G}_2(\mathbb{Q}_p)$$

be a continuous homomorphism such that $\mathrm{pr}_1(\pi(H)) \supset \mathbb{G}_1^+$ and $\mathrm{pr}_2(\pi(H)) \supset \mathbb{G}_2^+$. Then one of the following holds:

(1) $\pi(H) \supset \mathbb{G}_1^+ \times \mathbb{G}_2^+$.
(2) *There exists a morphism $\omega : \mathbb{G}_1 \to \mathbb{G}_2$ of algebraic groups, defined over $\mathbb{Q}_p$ such that $\pi(H) \subset Graph_\omega(\mathbb{Q}_p) = \{(g, \omega(g)) : g \in \mathbb{G}_1(\mathbb{Q}_p)\}$.*

Proof. The intersection $\pi(H) \cap (e \times \mathbb{G}_2(\mathbb{Q}_p))$ is normalized by $e \times \mathbb{G}_2^+$ and hence either contains $e \times \mathbb{G}_2^+$ or is trivial. In the first case we deduce that $\pi(H) \supset \mathbb{G}_1^+ \times \mathbb{G}_2^+$. Assume that the second case occurs. Let $K < H$ be a compact open subgroup. Since

H/K is countable it follows that $\mathrm{pr}_i(\pi(K))$ is a compact subgroup of $\mathbb{G}_i(\mathbb{Q}_p)$ which is of countable index in $\mathrm{pr}_i(\pi(H)) \supset \mathbb{G}_i^+$ and hence (as can be seen for example using the Baire category theorem) is open in $\mathbb{G}_i(\mathbb{Q}_p)$. Let $\mathfrak{G}_i = \mathrm{Lie}(\mathbb{G}_i)$ and since $\pi(K)$ is a closed subgroup of the p-adic Lie group $\mathbb{G}_1(\mathbb{Q}_p) \times \mathbb{G}_2(\mathbb{Q}_p)$, denote by $\mathcal{K}$ its Lie algebra. Since $\mathrm{pr}_i(\pi(K))$ is open in $\mathbb{G}_i(\mathbb{Q}_p)$ we have $\mathrm{pr}_i(\mathcal{K}) = \mathfrak{G}_i(\mathbb{Q}_p)$. Moreover since $\pi(K) \cap (e \times \mathbb{G}_2(\mathbb{Q}_p)) = \{e\}$, we have that

$$\mathcal{K} \cap (0 \oplus \mathfrak{G}_2(\mathbb{Q}_p)) = \{0\}. \tag{1.2.1}$$

Let then $\mathbb{L} \subset \mathbb{G}_1 \times \mathbb{G}_2$ denote the Zariski closure of $\pi(H)$. Since $\pi(H)$ commensurates $\pi(K)$, this implies that $\pi(H)$ acts on $\mathcal{K}$ by adjoint action and thus $\mathbb{L}(\mathbb{Q}_p)$ preserves $\mathcal{K}$. Together with (1.2.1) this implies that $\mathbb{L} \neq \mathbb{G}_1 \times \mathbb{G}_2$ and since $\mathrm{pr}_i(\mathbb{L}) = \mathbb{G}_i$, this implies that $\mathbb{L} \cap e \times \mathbb{G}_2 = \{e\}$ and hence $\mathbb{L}$ is the graph of a morphism of algebraic groups $\omega : \mathbb{G}_1 \to \mathbb{G}_2$ defined over $\mathbb{Q}_p$. □

A simple inductive argument gives:

Corollary 1.2.10. *Assume that $\mathbb{G}_1, \ldots, \mathbb{G}_n$ are connected $\mathbb{Q}_p$-almost simple of positive rank and $\pi : H \to \prod_{i=1}^n \mathbb{G}_i(\mathbb{Q}_p)$ is a continuous homomorphism so that*

(1) $\mathrm{pr}_i(H) \supset \mathbb{G}_i^+$.
(2) *$\pi(H)$ is Zariski dense.*

Then $\pi(H) \supset \prod_{i=1}^n \mathbb{G}_i^+$.

1.3. Arithmeticity

Our goal in this section is to prove Theorem 1.1.17. A cocompact lattice $\Gamma < \mathrm{Aut}T_1 \times \mathrm{Aut}T_2$ is said to be *LQP-sandwiched* if there exist closed non-discrete, locally quasiprimitive subgroups $H_i < \mathrm{Aut}T_i$ such that

$$H_i^{(\infty)} < \overline{\mathrm{pr}_i(\Gamma)} < H_i$$

Observe that since $H_i^{(\infty)}$ has no proper closed subgroup of finite index it follows that every finite index subgroup of Γ is also LQP-sandwiched.

Recall

Proposition 1.3.1 ([BM00b, Cor. 3.3]). *Let $\Gamma < \mathrm{Aut}T_1 \times \mathrm{Aut}T_2$ be a cocompact lattice which is LQP-sandwiched. Then*

(1) *Its abelianization Γ_{ab} is finite.*
(2) *For any unitary representation $\omega : \Gamma \to \mathrm{U}(n)$, $H^1(\Gamma, \omega) = 0$.*

Combining Theorem 1.1.3 and Proposition 1.3.1 we obtain:

Corollary 1.3.2. *Let $\Gamma < \mathrm{Aut}T_1 \times \mathrm{Aut}T_2$ be a LQP-sandwiched lattice. Then:*

(1) *Any homomorphism $\pi : \Gamma \to \mathrm{GL}(n, \mathbb{C})$ has bounded image.*
(2) *Let k be a field of characteristic zero and $\pi : \Gamma \to \mathrm{GL}(n, k)$ a homomorphism, then the Zariski closure $\overline{\pi(\Gamma)}^Z$ is semisimple.*

Proof. In (2) we may replace the field k by $\mathbb{C}$ and we will prove both (1) and (2) simultaneously. Let $\mathbb{L}$ be the connected component of the Zariski closure of the image of Γ and let $\Gamma' < \Gamma$ be a subgroup of finite index with $\overline{\pi(\Gamma')}^Z = \mathbb{L}$. Since Γ'_{ab} is finite and its image in $\mathbb{L}/[\mathbb{L}, \mathbb{L}]$ is Zariski dense, the latter is trivial and hence the radical $\mathrm{Rad}(\mathbb{L})$ coincides with the unipotent radical $\mathrm{Rad}_u(\mathbb{L})$. Let $\mathbb{S} = \mathbb{L}/\mathrm{Rad}(\mathbb{L})$ and $\overline{\pi} : \Gamma' \to \mathbb{S}$ be the composition of π with the canonical projection $\mathbb{L} \to \mathbb{S}$. If $\overline{\pi}(\Gamma)$ were unbounded then there would be a simple quotient $\mathbb{S} \to \mathbb{S}_1$ such that composing with $\overline{\pi}$ we would have a homomorphism $\pi_1 : \Gamma' \to \mathbb{S}_1$ with Zariski dense unbounded image. Setting $P_i = \overline{\mathrm{pr}_i(\Gamma')}$, we are in the setting of Theorem 1.1.3 and π_1 extends to a continuous homomorphism factoring via, say, P_1. Let $\pi_{1,ext} : P_1 \to \mathbb{S}_1$ be this continuous homomorphism. Since P_1 is totally disconnected, $\ker \pi_{1,ext} \lhd P_1$ is open. Since $P_1 \supset H_1{}^{(\infty)}$, we have $\ker \pi_{1,ext} \supset H_1{}^{(\infty)}$ (Proposition 1.2.4.2) and hence the image of $\pi_{1,ext}$ is compact, implying that $\pi_1(\Gamma')$ is bounded, a contradiction. Consider now $\mathbb{L}/[\mathrm{Rad}_u(\mathbb{L}), \mathrm{Rad}_u(\mathbb{L})] = \mathbb{S} \cdot V$, the semidirect product of $\mathbb{S}$ with a vector space V. The corresponding homomorphism $\Gamma' \to \mathbb{S} \cdot V$ has the form $\gamma \mapsto \overline{\pi}(\gamma) \cdot c(\gamma)$, where $\overline{\pi}(\Gamma')$ is bounded and $c : \Gamma \to V$ is a 1-cocycle with values in a bounded representation of Γ'. It follows from Prop 1.3.1.2 that this cocycle is trivial and hence that the image of Γ' in $\mathbb{S} \cdot V$ is contained in a conjugate of $\mathbb{S}$. Since the image is Zariski dense it follows that $V = 0$. I.e., $\mathrm{Rad}_u(\mathbb{L}) = [\mathrm{Rad}_u(\mathbb{L}), \mathrm{Rad}_u(\mathbb{L})]$ and hence we have $\mathrm{Rad}_u(\mathbb{L}) = 0$. □

We introduce the following terminology: Let k be a local field. A *k-triple* is a triple $(\Gamma', \pi, \mathbb{H})$ where $\Gamma' < \Gamma$ is a subgroup of finite index, $\mathbb{H}$ is a connected k-simple adjoint group of positive dimension and $\pi : \Gamma' \to \mathbb{H}(k)$ is a homomorphism with Zariski dense image. We say that it is of *unbounded type* if $\pi(\Gamma')$ is unbounded in $\mathbb{H}(k)$.

Lemma 1.3.3. *Let $(\Gamma', \pi, \mathbb{H})$ be a k-triple. Then there exist a finite extension $\mathbb{Q} \subset K \subset k$ of $\mathbb{Q}$ and a K-structure on $\mathbb{H}$ so that $\pi(\Gamma') \subset \mathbb{H}(K)$.*

Proof. Consider $\mathbb{H}$ as a subgroup of $\mathrm{GL}(n, k)$ via the adjoint representation, fix an embedding $\iota : k \to \mathbb{C}$ and let us abuse notation and denote by ι also the induced group homomorphism $\iota : \mathrm{GL}(n, k) \to \mathrm{GL}(n, \mathbb{C})$. The homomorphism $\iota \circ \pi : \Gamma' \to \mathrm{GL}(n, \mathbb{C})$ as well as any twist of it by an automorphism of $\mathbb{C}$ have bounded image (Corollary 1.3.2). It follows that for any $\gamma \in \Gamma'$ $\mathrm{tr}(\iota \circ \pi(\gamma)) = \iota(\mathrm{tr}\, \pi(\gamma))$ is an algebraic number. Since ι is the identity on $\mathbb{Q}$, it follows that $\mathrm{tr}\, \pi(\gamma)$ is in $\overline{\mathbb{Q}} \cap k$ for any $\gamma \in \Gamma'$. Since Γ' is finitely generated it follows that the field generated by all these traces is a finite extension of $\mathbb{Q}$ which we denote by K. It follows now by a standard argument using the Zariski density of $\pi(\Gamma')$ that there is a faithful k-rational representation $\rho : \mathbb{H} \to \mathrm{GL}(V)$ such that $\rho(\mathbb{H})$ is defined over K and $\rho(\pi(\Gamma')) \subset \rho(\mathbb{H})(K)$. □

Lemma 1.3.4. *Assume that there is a representation $\rho : \Gamma \to \mathrm{GL}(n, \mathbb{C})$ with infinite image. Then there exists a $\mathbb{Q}_p$-triple of unbounded type.*

Proof. By Corollary 1.3.2.2 the Zariski closure $\overline{\rho(\Gamma)}^Z$ is semisimple. Passing to a simple quotient we get a $\mathbb{C}$-triple $(\Gamma', \pi, \mathbb{H})$. Endow $\mathbb{H}$ with the K-structure given by Lemma 1.3.3. Then in particular $\pi(\Gamma') \subset \mathbb{H}(K)$. Let $\mathbb{L} = \mathrm{Res}_{K/\mathbb{Q}}\mathbb{H}$ and let

$\Delta : \mathbb{H}(K) \to \mathbb{L}(\mathbb{Q})$ be the "diagonal" isomorphism. Let S be the set of primes p for which $\Delta(\pi(\Gamma'))$ is unbounded in $\mathbb{L}(\mathbb{Q}_p)$. Using Corollary 1.3.2.2 as above it suffices to show that S is non-empty. Assume that $S = \emptyset$; then $\Delta(\pi(\Gamma'))$ is up to finite index, contained in $\mathbb{L}(\mathbb{Z})$, hence discrete. On the other hand $\Delta(\pi(\Gamma')) \subset \mathbb{L}(\mathbb{R})$ is bounded (Corollary 1.3.2.1) which implies that $\Delta(\pi(\Gamma'))$ is finite, contradicting the fact that $\pi(\Gamma')$ is Zariski dense in the non-trivial connected group $\mathbb{H}$. □

1.3.1. *Proof of the arithmeticity theorem 1.1.17*

Let P_Γ denote the set of primes p for which there exists a $\mathbb{Q}_p$-triple of unbounded type. According to Lemma 1.3.4, P_Γ is non-void. On the other hand let $(\Gamma', \pi, \mathbb{H})$ be a $\mathbb{Q}_p$-triple of unbounded type and denote $L_i = \overline{\mathrm{pr}_i(\Gamma')}$. Theorem 1.1.3 implies that π extends continuously to $L_1 \times L_2$ factoring via, say, L_1. Thus we obtain a continuous unbounded linear representation $L_1 \to \mathbb{H}(\mathbb{Q}_p)$. By Theorem 1.2.8 it follows that L_1 is $\mathbb{Q}_p$-analytic. This implies that $|P_\Gamma| \in \{1, 2\}$. Since the case where $|P_\Gamma| = 1$ is slightly more involved we shall describe the argument for it (leaving the other case to the reader) and assume henceforth that $|P_\Gamma| = 1$.

Claim 1.3.5. Given any $\mathbb{Q}_p$-triple $(\Gamma', \pi, \mathbb{H})$ of unbounded type. There are $\mathbb{Q}_p$-triples $(\Gamma'', \pi_i, \mathbb{H}_i)$, $1 \le i \le n$, of unbounded type such that $\Gamma'' < \Gamma'$ is of finite index, $\mathbb{H}_1 = \mathbb{H}$, $\pi_1 = \pi_{|_{\Gamma''}}$ and the product homomorphism $\prod \pi_i : \Gamma'' \to \prod_{i=1}^{n} \mathbb{H}_i(\mathbb{Q}_p)$ has Zariski dense discrete image.

Endow $\mathbb{H}$ with the K-structure given by lemma 1.3.3; $\mathbb{Q} \subset K \subset \mathbb{Q}_p$. Let Δ : $\mathbb{H}(K) \to \mathrm{Res}_{K/\mathbb{Q}}\mathbb{H}(\mathbb{Q})$ be the diagonal isomorphism, $\mathbb{L}$ the connected component of the Zariski closure of the image of Γ' and $\Gamma'' = (\Delta \circ \pi)^{-1}(\mathbb{L}(\mathbb{Q}))$. Then $\mathbb{L}$ is defined over $\mathbb{Q}$, semisimple (see Corollary 1.3.2) and Γ'' is of finite index in Γ'. Since p is the only prime for which $\Delta \circ \pi(\Gamma'')$ is unbounded in $\mathbb{L}(\mathbb{Q}_p)$, we deduce that, up to a subgroup of finite index, $\Delta\pi(\Gamma'')$ is contained in $\mathbb{L}(\mathbb{Z}[1/p])$; since moreover the image in $\mathbb{L}(\mathbb{R})$ if $\Delta(\pi(\Gamma''))$ is bounded (Corollary 1.3.2) we conclude that $\Delta(\pi(\Gamma''))$ is discrete in $\mathbb{L}(\mathbb{Q}_p)$. If one lets now $\mathbb{H}_1, \dots, \mathbb{H}_n$ be the $\mathbb{Q}_p$-simple adjoint quotients $p_i : \mathbb{L} \to \mathbb{H}_i$ of $\mathbb{L}$ such that $(\Gamma'', \pi_i, \mathbb{H}_i)$, with $\pi_i = p_i \circ \Delta \circ \pi$, is of unbounded type, then it is an easy verification that those fulfill the claim.

Let now $(\Gamma', \pi, \mathbb{H})$ be a $\mathbb{Q}_p$-triple of unbounded type and let $(\Gamma'', \pi, \mathbb{H}_i)$ be the triples given by the above claim. Each $\pi_i : \Gamma'' \to \mathbb{H}_i(\mathbb{Q}_p)$ extends continuously to $L_1 \times L_2$ (where $L_i = \overline{\mathrm{pr}_i(\Gamma'')}$) factoring via L_1 or L_2. Without loss of generality, assume that for $1 \le i \le r$, the extension factors via L_1 and let $\overline{\pi}_i : L_1 \to \mathbb{H}_i(\mathbb{Q}_p)$ be this continuous homomorphism and for $r+1 \le i \le n$ it factors via L_2 giving rise to a $\overline{\pi}_i : L_2 \to \mathbb{H}_i(\mathbb{Q}_p)$. Let $\alpha_1 = \prod_{i=1}^{r} \overline{\pi}_i$, $\alpha_2 = \prod_{i=r+1}^{n} \overline{\pi}_i$. It follows from Lemma 1.2.5.2 and Lemma 1.2.10 that $\alpha(H_1{}^{(\infty)}) = \prod_{i=1}^{r} \mathbb{H}_i^+$ and $\alpha_2(H_2{}^{(\infty)}) = \prod_{i=r+1}^{n} \mathbb{H}_i^+$.

Claim 1.3.6. $r < n$

Indeed if $r = n$ then the continuous extension of $\prod_{i=1}^{n} \pi_i : \Gamma'' \to \prod_{i=1}^{n} \mathbb{H}_i(\mathbb{Q}_p)$ to $L_1 \times L_2$ factors via the projection to L_1, the extension being given by $(\ell_1, \ell_2) \mapsto$

$\alpha_1(\ell_1)$. Using Claim 1.3.5 this implies that $\alpha_1(\mathrm{pr}_1(\Gamma''))$ is discrete, in particular closed and hence contains $\alpha_1(\overline{\mathrm{pr}_1(\Gamma'')}) = \alpha_1(L_1) \supset \prod_{i=1}^{n} \mathbb{H}_i^+$ which is impossible.

Thus $r < n$ and it follows from Theorem 1.2.8 that both H_1 and H_2 are p-adic analytic groups. Let $\mathfrak{G}_i$, $\mathbb{G}_i$ and $\mathrm{Ad}_i : H_i \to \mathrm{Aut}(\mathfrak{G}_i^{n_i})$ be as in Theorem 1.2.8. Let us switch notation and denote $\mathfrak{F}_i = \mathfrak{G}_i$, $\mathbb{F}_i = \mathbb{G}_i$. We have and

$$\mathrm{Ad} = \mathrm{Ad}_1 \times \mathrm{Ad}_2 : H_1 \times H_2 \to \mathrm{Aut}(\mathfrak{G}_1^{n_1}) \times \mathrm{Aut}(\mathfrak{G}_2^{n_2})$$
$$H_1{}^{(\infty)} \times H_2{}^{(\infty)} \twoheadrightarrow (\mathbb{G}_1^+)^{n_1} \times (\mathbb{G}_2^+)^{n_2}$$

Claim 1.3.7. $\mathrm{Ad}(\Gamma)$ is discrete.

Since $H_i{}^{(\infty)}$ is of finite index in H_i (Theorem 1.2.8.1) we may replace Γ by $\Gamma' = \Gamma \cap (H_1{}^{(\infty)} \times H_2{}^{(\infty)})$. Let $\mathbb{H}$ be any factor of $\mathbb{G}_1^{n_1}$, and let π be the composition of Ad with the projection on $\mathbb{H}$. Then $(\Gamma', \pi, \mathbb{H})$ is a $\mathbb{Q}_p$-triple of unbounded type to which Claim 1.3.5 and the subsequent construction applies. In particular we obtain

$$\alpha_1 : H_1{}^{(\infty)} \to \prod_{i=1}^{r} \mathbb{H}_i, \quad \alpha_1(H_1{}^{(\infty)}) = \prod_{i=1}^{r} \mathbb{H}_i^+,$$
$$\alpha_2 : H_2{}^{(\infty)} \to \prod_{i=r+1}^{n} \mathbb{H}_i, \quad \alpha_2(H_2{}^{(\infty)}) = \prod_{i=r+1}^{n} \mathbb{H}_i^+.$$

Observe that $\mathrm{QZ}(H_i{}^{(\infty)}) \subset \ker \alpha_i$. This follows, for example, from the fact that each $\mathbb{H}_i^+$ is an adjoint group. Using Theorem 1.2.8.3 we conclude that the maps α_i factor via $\mathrm{Ad}_i : H_i{}^{(\infty)} \to G_i^{n-i}$. I.e., the re are quotient maps $q_1 : \mathbb{G}_1^{n_1} \to \prod_{i=1}^{r} \mathbb{H}_i$, $q_2 : \mathbb{G}_2^{n_2} \to \prod_{i=r+1}^{n} \mathbb{H}_i$ such that

$$\begin{array}{ccc} H_1^{(\infty)} & \xrightarrow{\ \mathrm{Ad}_1\ } & \mathbb{G}_1^{n_1} \\ & \searrow^{\alpha_1} & \downarrow q_1 \\ & & \prod_{i=1}^{r} \mathbb{H}_i \end{array}$$

and

$$\begin{array}{ccc} H_2^{(\infty)} & \xrightarrow{\ \mathrm{Ad}_2\ } & \mathbb{G}_2^{n_2} \\ & \searrow^{\alpha_2} & \downarrow q_2 \\ & & \prod_{i=r+1}^{n} \mathbb{H}_i \end{array}$$

commute. In particular $(\alpha_1 \times \alpha_2)(\Gamma') = (q_1 \times q_2)(\mathrm{Ad}(\Gamma'))$ is discrete.

Since this applies to any factor of $\mathbb{G}_1^{n_1}$ and similarly for $\mathbb{G}_2^{n_2}$, we deduce that $\mathrm{Ad}(\Gamma)$ is discrete. We deduce from the facts that $\mathrm{Ad} : H_1{}^{(\infty)} \times H_2{}^{(\infty)} \to (\mathbb{G}_1^+)^{n_1} \times (\mathbb{G}_2^+)^{n_2}$ is surjective, Γ' is a cocompact lattice and $\mathrm{Ad}(\Gamma')$ is discrete that $\mathrm{Ad}(\Gamma')$ is a cocompact lattice in $\mathbb{G}_1(\mathbb{Q}_p)^{n_1} \times \mathbb{G}_2(\mathbb{Q}_p)^{n_2}$ for which the closure of the projection on the i'th factor contains $(\mathbb{G}_i^+)^{n_i}$, for $i = 1, 2$. This implies that Γ is virtually a product of irreducible lattices in semisimple (not simple) Lie groups of rank at least 2 This implies that $\mathrm{Ad}(\Gamma')$ is an arithmetic lattice and completes the proof in the case $P_\Gamma = \{p\}$. □

Proof of Corollary 1.1.18. 1 $\Rightarrow$ 2 is just the well-known fact that finitely generated linear groups are residually finite.

2 $\Rightarrow$ 3. Since Γ is residually finite, Proposition 1.1.11.1 implies that $\Lambda_i = \{e\}$, $i = 1, 2$ and hence (by Theorem 1.1.17) we have $\mathrm{QZ}(H_i) = \ker \mathrm{Ad}_i = \{e\}$. Thus $\mathrm{Ad}_i : H_i^{(\infty)} \to \mathbb{G}_i^+$ is a topological isomorphism. By considering the action on T_i (via $(\mathrm{Ad}_i)^{-1}$) of a split Cartan subgroup of $\mathbb{G}_i$, one deduces that $\mathbb{G}_i$ has $\mathbb{Q}_{p_i}$-rank 1.

3 $\Rightarrow$ 1. The locally primitive group H_i acts via the adjoint map Ad_i on the Bruhat–Tits tree Δ_i associated with $\mathbb{G}_i$. By Lemma 1.4.7 it follows that this action is implemented via an isometry between the geometric realizations $|T_i|$ and $|\Delta_i|$. This shows that Ad_i is injective and hence Γ is linear. □

1.3.2. *Proof of theorem 1.1.12*

Theorem 1.1.12 follows from Theorem 1.1.5 and the following:

Lemma 1.3.8. *Let $\Gamma < \mathrm{Aut}T_1 \times \mathrm{Aut}T_2$ be a cocompact lattice such that $H_i = \overline{\mathrm{pr}_i(\Gamma)}$ are non-discrete, locally quasiprimitive. Let $N \lhd \Gamma$ be a normal subgroup. Then either*

1. $N \subset \mathrm{QZ}(H_1) \times \mathrm{QZ}(H_2)$, *or*
2. $\overline{\mathrm{pr}_i(N)} \supset H_i^{(\infty)}$.

Proof. The closure $\overline{\mathrm{pr}_i(N)}$ is a normal subgroup of H_i and hence (by Theorem 1.1.7) either contains $H_i^{(\infty)}$ or is contained in $\mathrm{QZ}(H_i)$. Hence assume

$$\mathrm{pr}_1(N) \subset \mathrm{QZ}(H_1); \tag{1.3.1}$$

we have to show

$$\mathrm{pr}_2(N) \subset \mathrm{QZ}(H_2). \tag{1.3.2}$$

Assume by contradiction that (1.3.2) does not hold. Let $\Lambda_i = \Gamma \cap \mathrm{Aut}T_i$ (where we slightly abuse notation). Since $\overline{\mathrm{pr}_2(N)}$ is normal in $\overline{\mathrm{pr}_2(\Gamma)} = H_2$ we have by Theorem 1.1.7 that

$$\overline{\mathrm{pr}_2(N)} \supset H_2^{(\infty)} \tag{1.3.3}$$

But then (using 1.3.2 in [BM00a]) there is a finitely generated group $L \subset N$ such that $\mathrm{pr}_2(L)$ acts co-finitely on the tree T_2. Observe that $(\mathrm{pr}_i(L) \cap \Lambda_i) \lhd \mathrm{pr}_i(L)$ and that $\mathrm{pr}_1(L)/(\mathrm{pr}_1(L) \cap \Lambda_1)$ is isomorphic to $\mathrm{pr}_2(L)/(\mathrm{pr}_2(L) \cap \Lambda_2)$, indeed both are isomorphic to $L/((\ker \mathrm{pr}_1 |_L) \cdot (\ker \mathrm{pr}_2 |_L))$. Since $\mathrm{pr}_1(L)/(\mathrm{pr}_1(L) \cap \Lambda_1)$ is isomorphic to a subgroup of $\mathrm{QZ}(H_1)/\Lambda_1$ it is locally finite (Proposition 1.1.11.2). Hence also the finitely generated group $\mathrm{pr}_2(L)/(\mathrm{pr}_2(L) \cap \Lambda_2)$ is locally finite and hence finite. But then $\mathrm{pr}_2(L) \cap \Lambda_2$ would act on T_2 with a finite quotient. Hence Λ_2 and $\mathrm{QZ}(H_2)$ will be cocompact in H_2, which contradicts Theorem 1.1.7, we conclude that $\mathrm{pr}_2(N) \subset \mathrm{QZ}(H_2)$. □

1.4. Geometric rigidity

1.4.1.

Terminology: our trees will be without leaves. Given a tree T let us denote by $|T|$ its geometric realization which is a CAT(-1) space, endowed with the structure of a

1-dimensional simplicial complex. Let $|T|^0$ denote the set of 0-cells (corresponding to the vertices of T) and every 1-cell of $|T|$ is isometric to $[0, 1]$. A group action on a tree T is called c-minimal if there is no proper invariant subtree.

Theorem 1.4.1. *Let $\Gamma < \mathrm{Aut}T_1 \times \mathrm{Aut}T_2$ be a cocompact lattice such that $H_i = \overline{\mathrm{pr}_i(\Gamma)}$ are locally primitive and non-discrete. Let $\Gamma' < \mathrm{Aut}T_1' \times \mathrm{Aut}T_2'$ be such that $\Gamma' \backslash (T_1' \times T_2')$ is finite and let $\pi : \Gamma \to \Gamma'$ be a surjective homomorphism. Then up to re-scaling metrics and exchanging the factors there are isometries $t_i : |T_i| \to |T_i'|$ such that $t := t_1 \times t_2$ induces the homomorphism π, in particular π is an isomorphism.*

Remark 1.4.2. If we assume that π is an isomorphism we do not need to impose the condition that H_i is non-discrete.

Corollary 1.4.3. *Let $\Gamma < \mathrm{Aut}T_1 \times \mathrm{Aut}T_2$ be a cocompact lattice such that each $H_i = \overline{\mathrm{pr}_i(\Gamma)}$ is locally primitive and $X_\Gamma = \Gamma \backslash (T_1 \times T_2)$ be the quotient square complex. Then*

$$\mathrm{Out}(\Gamma) \cong \mathrm{Aut}X_\Gamma$$

and hence is finite.

The above results follow from a general result describing all non-elementary actions of Γ on a tree:

Theorem 1.4.4. *Let $\Gamma < \mathrm{Aut}T_1 \times \mathrm{Aut}T_2$ be a uniform lattice such that each $H_i = \overline{\mathrm{pr}_i(\Gamma)}$ is locally primitive and $\pi : \Gamma \to \mathrm{Aut}T$ be an action of Γ on a countable tree T such that it is non-elementary and c-minimal. Then π extends continuously to $H_1 \times H_2$, factoring via one H_i and the continuous homomorphism $\pi : H_i \to \mathrm{Aut}T$ obtained is realized by an isometry $|T_i| \to |T|$.*

Let us mention one more corollary:

Corollary 1.4.5. *Let $\Gamma < \mathrm{Aut}T_1 \times \mathrm{Aut}T_2$ be a cocompact lattice and assume that $H_i = \overline{\mathrm{pr}_i(\Gamma)}$ is locally primitive. Let $\Lambda_1 = \Gamma \cap (\mathrm{Aut}T_1 \times e)$ and $\Lambda_2 = \Gamma \cap (e \times \mathrm{Aut}T_2)$. If both $\Lambda_1 \neq \{e\}$ and $\Lambda_2 \neq \{e\}$, then $\Gamma/(\Lambda_1 \cdot \Lambda_2)$ has property FA. Namely any action of $\Gamma/(\Lambda_1 \cdot \Lambda_2)$ on a tree has a fixed point.*

1.4.2.

In this subsection we shall prove Theorem 1.4.4. We will need the following lemmas:

Lemma 1.4.6. *Let $T = (V, E)$ be a locally finite tree, $H < \mathrm{Aut}T$ a closed locally primitive subgroup, X a complete CAT(0)-space and $H \times X \to X$ a continuous action with unbounded orbits. Then after re-scaling of the metric on X there is a continuous H-equivariant map $\alpha : |T| \to X$ whose restriction to each 1-cell is isometric and whose restriction to the star of each 0-cell of $|T|$ is injective.*

Proof. Subdividing T once we may assume that H does not contain inversions and hence $H = \langle H_\alpha \cup H_\beta \rangle$ where α, β are any pair of adjacent vertices. Since for every vertex v the subgroup $\pi(H_v)$ has bounded orbits in X, it follows that $X^{\pi(H_v)} \neq \emptyset$

and we may pick $x_\alpha \in X^{\pi(H_\alpha)}$, $x_\beta \in X^{\pi(H_\beta)}$ and define $\varphi : V \to X$ by $\varphi(h\alpha) = \pi(h)x_\alpha$, $\varphi(h\beta) = \pi(h)x_\beta$, $h \in H$, thus obtaining an H-equivariant map from V to X. Pick $v \in V$ and let $\{x_1, \dots, x_k\}$ be the set of vertices adjacent to v. We show that $\varphi : \{v, x_1, \dots, x_k\} \to X$ is injective. If $\varphi(v) = \varphi(x_1)$, say, then $\pi(H_v)$ and $\pi(H_{x_1})$ both fix $b := \varphi(v) = \varphi(x_1)$ and hence $\pi(H)$ fixes b, a contradiction. The map $\varphi : \{x_1, \dots, x_k\} \to \{\varphi(x_1), \dots, \varphi(x_k)\}$ is H_v-equivariant and hence since the H_v action is primitive, it is either injective or constant. In the latter case, let $b \in X$ denote this constant value of the image. Then each of the subgroups $\pi(H_{x_i})$, $1 \le i \le k$ fixes b and so does the subgroup $N = \langle \cup H_{x_i} \rangle$ generated by them. Notice that $\pi(H_v)$ normalizes N and acts with bounded orbits on the subset $X^N \subset X$ of points fixed by N. Since $X^N \ni b$ is non-empty it follows that there is a point fixed by $\pi(H_v \cdot N) = \pi(H)$, a contradiction. In the sequel we pick x_α and x_β (and hence $\varphi()$ such that $[x_\alpha, x_\beta] \cap X^{\pi(H_\alpha)} = \{x_\alpha\}$ and $[x_\alpha, x_\beta] \cap X^{\pi(H_\beta)} = \{x_\beta\}$. Note that such a choice is clearly possible. Let $|\varphi| : |T| \to X$ be the geodesic extension of φ to $|T|$. Let $T_v = \cup_{i=1}^k [v, x_i]$; we have to show that $|\varphi|$ is injective on T_v. We say that $x_i \sim x_j$ if $[\varphi(v), \varphi(x_i)] \cap [\varphi(v), \varphi(x_j)] = [\varphi(v), q]$ with $q \neq \varphi(v)$. Clearly this is an H_v-invariant equivalence relation and hence is either separating points, i.e., $[\varphi(v), \varphi(x_i)] \cap [\varphi(v), \varphi(x_j)] = \{\varphi(v)\}$, $\forall i \neq j$, or consists of one equivalence class, in which case we have $\cap_{i=1}^k [\varphi(v), \varphi(x_i)] = [\varphi(v), q]$ with $q \neq \varphi(v)$. In this case however the point q is $\pi(H_v)$-fixed, contradicting the construction of φ. It follows that $|\varphi|_{|T_v}$ is injective on T_v. □

Lemma 1.4.7. *Let T and H be as in Lemma 1.4.6. Let T' be a countable tree, with a continuous H-action $\pi : H \to \mathrm{Aut}T'$ which is c-minimal and has unbounded orbits. Then $\pi : H \to \mathrm{Aut}T'$ is realized by an isometry $|T| \to |T'|$ (up to possible re-scaling of the distance on T').*

Proof. Apply Lemma 1.4.6 to $X = |T'|$ and let $\alpha : |T| \to |T'|$ be the H-equivariant map given by Lemma 1.4.6. We may assume that all the vertices of T have degree at least 3. Together with Lemma 1.4.6 this implies that $\alpha(|T|^0) \subset |T'|^0$ hence α is locally distance preserving. Observe that $\alpha(|T|) \subset |T'|$ is an H-invariant subtree and hence using the c-minimality of the action $\alpha(|T|) = |T'|$. A surjective map which is uniformly locally isometric is necessarily a covering map. Hence α is an isometry. □

Theorem 1.4.4 is an immediate consequence of Lemma 1.4.7 and Theorem 1.1.4 of Monod and Shalom.

1.4.3.

Here we complete the remaining assertions. To establish Theorem 1.4.1 consider $\pi_i := \mathrm{pr}_i \circ \pi : \Gamma \to \mathrm{Aut}T_i'$. These are non-elementary, c-minimal actions hence by Theorem 1.4.4 extend continuously to $\tilde{\pi}_i : H_1 \times H_2 \to \mathrm{Aut}T_i'$ factoring via one of the factors. Let us denote $\tilde{\pi} = \tilde{\pi}_1 \times \tilde{\pi}_2$. If both $\tilde{\pi}_1$ and $\tilde{\pi}_2$ factored via, say, H_1 then we will have (abusing notation) that $\tilde{\pi}(H_1 \times H_2) = \tilde{\pi}(H_1) \subset \overline{\tilde{\pi}(\mathrm{pr}_1(\Gamma)} = \Gamma'$. It follows that the image of H_1 is countable and hence must be finite (since the kernel would be an open normal subgroup of a locally primitive group), which is impossible. We conclude that, after possible exchanging the indexes, the map $\tilde{\pi}_i$ factors via H_i, and the resulting homomorphism $\overline{\pi_i} : H_i \to \mathrm{Aut}T_i'$ is realized (see Lemma 1.4.7) by an isometry $t_i : |T_i| \to |T_i'|$ and hence the homomorphism π is realized by the isometry $t = t_1 \times t_2$. □

The remaining Corollaries 1.4.3 and 1.4.5 follow easily.

References

[BG02] Nadia Benakli and Yair Glasner, Automorphism groups of trees acting locally with affine permutations, *Geom. Dedicata*, **89** (2002) 1–24. MR2003a:20043.

[BL01] Hyman Bass and Alexander Lubotzky, Tree lattices, Progress in Mathematics, vol. 176, Birkhäuser Boston Inc., Boston, MA, 2001, With appendices by H. Bass, L. Carbone, Lubotzky, G. Rosenberg and J. Tits. MR2001k:20056.

[BM96] Marc Burger and Shahar Mozes, CAT(-1)-spaces, divergence groups and their commensurators, *J. Amer. Math. Soc.*, **9** (1996) no. 1, 57–93. MR96c:20065.

[BM97] Marc Burger and Shahar Mozes, Finitely presented simple groups and products of trees, *C. R. Acad. Sci. Paris Sér. I Math.*, **324** (1997) no. 7, 747–752. MR98g:20041.

[BM00a] Marc Burger and Shahar Mozes, Groups acting on trees: from local to global structure, *Inst. Hautes Études Sci. Publ. Math.*, (2000) no. 92, 113–150 (2001). MR2002i:20041.

[BM00b] Marc Burger and Shahar Mozes, Lattices in product of trees, *Inst. Hautes Études Sci. Publ. Math.*, (2000) no. 92, 151–194 (2001). MR2002i:20042.

[BnK70] I. N. Bernšteĭ n and D. A. Každan, The one-dimensional cohomology of discrete subgroups, *Funkcional. Anal. i Priložen.*, **4** (1970) no. 1, 1–5. MR44 #1768.

[Bon] P. Bonvin, Strong boundaries and commensurator superrigidity, Appendix to B. Remy, Topological simplicity, commensurator super-rigidity and non-linearity of Kac-Moody groups, prépublication de l'Institut de Fourier (2002).

[BS03] Uri Bader and Yehuda Shalom, Factor and normal subgroup theorems for lattices in products of groups (2003).

[Bur95] M. Burger, Rigidity properties of group actions on CAT(0)-spaces, Proceedings of the International Congress of Mathematicians, Vol. 1, 2 (Zürich, 1994) (Basel), Birkhäuser, (1995) pp. 761–769. MR97j:20033.

[Cam95] Peter J. Cameron, Permutation groups, Handbook of combinatorics, Vol. 1, 2, Elsevier, Amsterdam (1995) pp. 611–645. MR97e:20002.

[Dic22] L. E. Dickson, Arithmetic of quaternions, *Proc. London Math. Soc. (2)*, **20** (1922) 225–232.

[DM96] John D. Dixon and Brian Mortimer, Permutation groups, Graduate Texts in Mathematics, vol. 163, Springer-Verlag, New York (1996). MR98m:20003.

[Gao97] Yan Gao, Superrigidity for homomorphisms into isometry groups of CAT(−1) spaces, *Transform. Groups*, **2** (1997) no. 3, 289–323. MR98h:53085.

[GKM08] Tsachik Gelander, Anders Karlsson and Gregory A. Margulis, Superrigidity, generalized harmonic maps and uniformly convex spaces, *Geom. Funct. Anal.*, **17** (2008) no. 5, 1524–1550.

[Gla03] Yair Glasner, A two-dimensional version of the Goldschmidt-Sims conjecture, *J. Algebra*, **269** (2003) no. 2, 381–401. MR2 015 283.

[HM79] Roger E. Howe and Calvin C. Moore, Asymptotic properties of unitary representations, *J. Funct. Anal.*, **32** (1979) no. 1, 72–96. MR80g:22017

[Mar74] Gregory A. Margulis, Arithmeticity and finite-dimensional representations of uniform lattices, *Funkcional. Anal. i Priložen.*, **8** (1974) no. 3, 77–78. MR50 #10157.

[Mar79] Gregory A. Margulis, Finiteness of quotient groups of discrete subgroups, *Funktsional. Anal. i Prilozhen.*, **13** (1979) no. 3, 28–39. MR80k:22006.

[Mar81] Gregory A. Margulis, On the decomposition of discrete subgroups into amalgams, *Selecta Math. Soviet.*, **1** (1981) no. 2, 197–213. MR84j:22016.

[Mar91] Gregory A. Margulis, Discrete subgroups of semisimple Lie groups, Ergebnisse der Mathematik und ihrer Grenzgebiete (3) [Results in Mathematics and Related Areas (3)], vol. 17, Springer-Verlag, Berlin (1991). MR92h:22021.

[Mar94] Gregory A. Margulis, Superrigidity of commensurable subgroups and generalized harmonic maps, unpublished (1994).

[Mon05] Nicolas Monod, Superrigidity for irreducible lattices and geometric splitting, *C. R. Math. Acad. Sci. Paris*, **340** (2005) no. 3, 185–190. MR2123025 (2005i: 22009).

[Mon06] Nicolas Monod, Superrigidity for irreducible lattices and geometric splitting, *J. Amer. Math. Soc.*, **19** (2006) no. 4, 781–814 (electronic). MR2219304 (2007b: 22025).

[Moo66] Calvin C. Moore, Ergodicitiy of flows on homogeneous spaces, *Amer. J. Math.*, **88** (1966) 154–178. MR33 #1409.

[Moz98] Shahar Mozes, *Products of trees, lattices and simple groups*, Proceedings of the International Congress of Mathematicians, Vol. II (Berlin, 1998), no. Extra Vol. II, 1998, pp. 571–582 (electronic). MR2000a:20056.

[MS02] Nicolas Monod and Yehuda Shalom, Cocycle superrigidity and bounded cohomology for negatively curved spaces (2002).

[Pra97] Cheryl E. Praeger, Finite quasiprimitive graphs, Surveys in combinatorics, 1997 (London), London Math. Soc. Lecture Note Ser., vol. 241, Cambridge Univ. Press, Cambridge (1997) pp. 65–85. MR99b:05076.

[Rat04] Diego Rattaggi, Computations in groups acting on a product of trees: normal subgroup structure and quaternion lattices, Ph.D. thesis, ETH Zürich (2004).

[Ser03] Jean-Pierre Serre, Trees, Springer Monographs in Mathematics, Springer-Verlag, Berlin, 2003, Translated from the French original by John Stillwell, Corrected 2nd printing of the 1980 English translation. MR2003m:20032

[Sha00] Yehuda Shalom, Rigidity of commensurators and irreducible lattices, *Invent. Math.*, **141** (2000) no. 1, 1–54. MR2001k:22022.

[Wis95] Dani Wise, A non-positively curved square complex with no finite covers (1995).

[Wis96] Dani Wise, Non-positively curved square complexes, aperiodic tilings and non-residually finite groups, Ph.D. thesis, Princeton (1996).

Proc. Inst. Wrk.shps. Int. Conf. – Essays in Geometric Group Theory No. 9, 2009, pp. 27–60.

Some Aspects of Profinite Group Theory

Dan Segal

All Souls College, Oxford OX1 4AL, UK
e-mail: dan.segal@all-souls.ox.ac.uk

2.1. Introduction

2.1.1. *Origins*

Profinite groups were introduced in number theory early in the last century. First of all, the group of p-adic integers $\mathbb{Z}_p$ appeared as a means for studying congruences: one can replace infinitely many congruences of the form

$$f(\mathbf{X}) \equiv 0 \,(\mathrm{mod}\, p^n)$$

by a single *equation*

$$f(\mathbf{X}) = 0$$

over $\mathbb{Z}_p$. There are two advantages to this approach. One is that we can do arithmetic in a nice integral domain of characteristic zero, instead of the messy finite rings $\mathbb{Z}/p^n\mathbb{Z}$. More importantly, though, from a methodological point of view, what we have here is a technology for replacing infinitely many hypotheses (about disparate small objects) with a single hypothesis (about one large object): the "large object" – the p-adic integers in this case – can then be studied by methods of algebra or arithmetic. This process of "mathematical reification" is of course quite traditional (as in the construction of the complex numbers), but is a particularly characteristic feature of 20th century mathematics (Hilbert space, representable functors, ...).

As a profinite group, of course, $\mathbb{Z}_p$ is rather trivial, and its main role in this context is as a *ring*. Profinite groups of (much) greater complexity were introduced by Krull. His insight was that the Galois group of an infinite algebraic Galois extension of fields is in a natural way a profinite group: it is a compact topological group, whose structure is completely determined by the *finite* Galois groups of all the finite Galois subextensions. This led to the elegant modern formulations of class field theory by Chevalley, Artin and Tate.

Later, Grothendieck introduced profinite groups into algebraic geometry, as the fundamental groups of schemes. I shall say no more about these topics, which are well beyond my competence: instead, I will concentrate on profinite groups as objects of study for group theorists. This is not going to be a comprehensive survey even of this limited subject: my intention is merely to point the reader to some areas where interesting developments have recently taken place, and that I happen to know something about (and the different amounts of space devoted to the various topics in no way reflects their relative importance).

I would like to acknowledge a special debt to both Alex Lubotzky and Avinoam Mann, from whom – way back in the last century – I learnt new ways to do group theory. Thanks are also due to Alex Lubotzky and Derek Holt for useful contributions to this article.

2.1.2. *"Profinite group theory"*

This phrase, like 'algebraic number theory', has a useful ambiguity (which is lost on translation into French); I intend the reader to keep in mind both meanings – 'the theory of profinite groups' and 'a profinite approach to group theory'.

A profinite group is what you get when you look at a (suitably coherent) collection of finite groups all at once. In this context, 'coherent' means that the groups in question form an *inverse system*: a family of finite groups (G_λ) indexed by a directed set Λ, and for each pair $\alpha, \beta \in \Lambda$ with $\alpha \leq \beta$ a homomorphism $\theta_{\beta\alpha} : G_\beta \to G_\alpha$. Whenever $\alpha \leq \beta \leq \gamma$ we require that $\theta_{\beta\alpha} \circ \theta_{\gamma\beta} = \theta_{\gamma\alpha}$, and each $\theta_{\alpha\alpha}$ is the identity automorphism. (To say that Λ is a directed set means that Λ is partially ordered and that for every pair $\alpha, \beta \in \Lambda$ there exists $\gamma \in \Lambda$ with $\gamma \geq \alpha$ and $\gamma \geq \beta$.) The *inverse limit* of this system, denoted

$$\lim_{\overleftarrow{\Lambda}} G_\lambda,$$

may be defined by a suitable universal property, or more concretely as a subgroup G of the Cartesian product of all the G_λ, as follows:

$$G = \left\{(g_\lambda) | \theta_{\beta\alpha}(g_\beta) = g_\alpha \text{ whenever } \beta > \alpha\right\} \leq \prod_{\lambda \in \Lambda} G_\lambda. \tag{2.1.1}$$

Thus G maps naturally into each of the finite groups G_λ (by projecting to a factor), and G is completely determined by the system $(G_\lambda)_{\lambda \in \Lambda}$; the homomorphisms $\theta_{\beta\alpha}$ are supposed to be included as part of the definition of the system. So far, we have done little more than introduce a notation. The key observation now is that G is in a natural way a *topological group*: giving each of the finite groups G_λ the discrete topology we endow $\prod_{\lambda \in \Lambda} G_\lambda$ with the product topology; instead of being discrete, this is a compact Hausdorff space, by Tychonoff's Theorem. It is easy to see that the inverse limit G is a closed subgroup, so in this way G becomes a *compact Hausdorff topological group*. For each $\lambda \in \Lambda$ the kernel K_λ of the projection $\pi_\lambda : G \to G_\lambda$ is an open normal subgroup of G, and the family $\{K_\lambda\}$ forms a base for the neighbourhoods of 1 in G. In most naturally-arising situations, the maps $\theta_{\beta\alpha}$ are all surjective, in which case one speaks of a *surjective inverse system*. The projections π_λ are then also surjective ([RZ2], Chapter 1), so the original groups G_λ appear as continuous finite images of G. To save repetition, I will make the blanket assumption that *all inverse systems under discussion are surjective*; this is no real loss of generality, since to the inverse system (G_λ) we may associate another one $(\pi_\lambda(G))$ which is surjective and has the same inverse limit G.

The bare algebraic structure of G may carry little information about the original system of finite groups, but in combination with the topology it closely reflects many properties of that system. Vaguely speaking, properties of the topological group G reflect *uniform* properties of the groups G_λ. For example, we find that G is finitely generated (as a *topological* group) if and only if there exists $d \in \mathbb{N}$ such that each of the groups G_λ can be generated by d elements; more subtle relationships of this kind will be discussed below.

Certain classes of profinite groups have special names: if the finite groups G_λ all belong to some class of groups $\mathcal{C}$ (and the inverse system is surjective), then $G = \varprojlim G_\lambda$ is called a pro-$\mathcal{C}$ group. When $\mathcal{C}$ is the class of finite p-groups for some prime p one calls G a pro-p group.

In practice, most of the questions studied in "profinite group theory" arise in one of the following contexts, which are not mutually exclusive.

(1) Questions about some naturally-defined family of finite groups, for example finite p-groups; see §2.5.
(2) Questions about infinite groups that can be approached through their profinite completions; this may be construed as a subcase of (1) where the family of finite groups consists of finite quotients of some fixed infinite group. See §§2.2, 2.4, 2.8.1.
(3) Questions about profinite groups as such; these may be analogues in the profinite category of familiar group-theoretic questions (§2.8.3, 2.8.2), they may arise from number theory and field theory via the Galois group (see [FJ], [B], [dSF]), or they may be a new kind of question specific to the profinite situation (§§2.6, 2.7). As we shall see, such investigations sometimes lead to new results about abstract groups, finite or infinite.

For definitions and the basic properties of profinite groups, consult [DDMS] Chapter 1, [W2] Chapter 1, or [RZ2] Chapter 2. Each of these books goes on to discuss various specific topics; some of them are mentioned below, but all are worth studying. Galois-theoretic applications of profinite groups are pursued at length in [FJ]. Various aspects of pro-p groups are discussed in detail in [NH], which includes a substantial list of open problems.

The first substantial treatment in book form of profinite groups was Serre's influential book [CG]: as the title suggests, this is primarily concerned with homological matters and is slanted towards number theory.

2.2. Local and global

An important strand in number theory is the investigation of so-called 'local-global' principles. A typical question is like this: suppose a certain Diophantine equation $f(\mathbf{X}) = c$ can be solved modulo m for every $m \in \mathbb{N}$, does it follow that $f(\mathbf{X}) = c$ has a solution in integers? The most famous example is the Hasse-Minkowski Theorem, which gives an affirmative answer for *rational* solutions, at least, when $f(\mathbf{X})$ is an indefinite quadratic form. A fruitful way of formulating such statements was introduced by Hasse (inspired by Hensel): instead of considering congruences one takes the (equivalent) hypothesis that $f(\mathbf{X}) = c$ is solvable in every p-adic field (if we call $\mathbb{R}$ the '∞-adic field' this also subsumes the condition that the quadratic form be indefinite). In this case, the equation is said to be solvable 'locally'; if this implies the existence of a rational solution one has a 'local-global' theorem, and the equation is said to satisfy the 'Hasse Principle'.

In the case of quadratic forms, the answer for *integral* solutions is a little more complicated: it may be that $f(\mathbf{X}) = c$ is not solvable in integers, but at least we can say that $f'(\mathbf{X}) = c$ is solvable where $f'(\mathbf{X})$ is one of *finitely many* quadratic forms, that constitute the *genus* of f.

What has this to do with group theory? Thinking of the above as the search for properties of $\mathbb{Z}$ that are determined by properties of the collection of finite rings $\mathbb{Z}/m\mathbb{Z}$, we can generalize as follows: to what extent are properties of an infinite group G determined by the finite quotient groups of G? This is a natural enough question in itself; it also has a further philosophical motivation, connected with *decision problems*. The point is that a 'local-global' theorem in group theory, as in number theory, often implies a corresponding decidability theorem. Rather than stating this as a formal metatheorem let me illustrate with an example which should make the idea clear. A group G is *conjugacy separable* if it has the following property: if $x, y \in G$ are such that $\pi(x)$ is conjugate to $\pi(y)$ in $\pi(G)$ for every homomorphism π from G to any finite group, then x and y are conjugate in G; this is the 'local-global' property of conjugacy in G. On the other hand, G has *solvable conjugacy problem* if there is a uniform algorithm that decides, given any two elements of G, whether or not they are conjugate in G.

Theorem 2.2.1. *Every finitely presented conjugacy separable group has solvable conjugacy problem.*

The algorithm consists of two procedures, run simultaneously. The first one lists all consequences of the relations in a given presentation of G, while the second one enumerates all homomorphisms π from G to finite groups, and for each such π lists the (finitely many) pairs of non-conjugate elements in $\pi(G)$. Now given x and $y \in G$, we run both procedures until *either* the first one spits out an equality $x^g = y$ *or* the second one spits out a pair $(\pi(x), \pi(y))$. In the first case we conclude that x and y are conjugate, in the second that they are not; the hypothesis that G is conjugacy separable ensures that one or other of the cases must arise.

Of course, no sane person would try to implement such a stupid algorithm; its interest is theoretical. It shows that combinatorial group theorists shouldn't waste their time trying to prove the unsolvability of the conjugacy problem in the case of conjugacy separable groups. The same applies to the *word* problem in *residually finite* groups: a group is residually finite if its subgroups of finite index intersect in $\{1\}$, which is equivalent to saying that any two distinct elements have distinct images in at least one finite quotient of the group – the local-global property for equality of elements (while the 'word problem' asks for an algorithm to determine equality of group elements, given as words on a fixed generating set).

As in number theory, there is a useful reformulation for group-theoretic 'local-global' questions. The family of all finite quotients of G naturally forms an inverse system of finite groups, with respect to the quotient maps

$$G/N \to G/M$$

where $N \leq M$ are normal subgroups of finite index in G. The inverse limit of this system is the *profinite completion* $\widehat{G}$ of G; and the question becomes: what properties of G are determined by properties of the profinite group $\widehat{G}$?

The family of quotient maps $G \to G/N$ induces a natural homomorphism $\iota : G \to \widehat{G}$. The kernel of ι is the *finite residual* $R(G)$ of G, which is the intersection of all subgroups of finite index in G. Evidently $\widehat{G} = \widehat{G/R(G)}$, so knowledge of $\widehat{G}$ will at best give us information about $G/R(G)$; thus it is sensible to restrict attention to groups G for which $R(G) = 1$, that is residually finite groups. If G is residually finite then the map ι is injective, and we use it to identify G with a subgroup of $\widehat{G}$. This amounts to identifying an element $g \in G$ with the 'diagonal' element

$$(gN)_{N \in \mathcal{N}} \in \widehat{G} \leq \prod_{N \in \mathcal{N}} G/N$$

where $\mathcal{N}$ is the family of all normal subgroups of finite index in G.

To say that G is residually finite, then, amounts to saying that two elements of G are equal if and only if they map to equal elements of $\widehat{G}$. Similarly, G is conjugacy separable if and only if for pairs of elements of G, conjugacy in $\widehat{G}$ implies conjugacy in G; this is equivalent to saying that *each conjugacy class* in G *is closed* in the profinite topology of G, that is, the topology induced from $\widehat{G}$, in which a base for the neighbourhoods of 1 in G is given by the family $\mathcal{N}$ of all normal subgroups of finite index in G. (Analogously, G is residually finite if and only if *points* are closed in G – in this context this is equivalent to the Hausdorff property.)

Well known classes of residually finite groups include the *free groups* and the *virtually polycyclic groups*. (A group is *virtually* P if it has a normal P-subgroup of finite index.) In fact, groups in these classes have many good local-global properties: in particular, they are

- *conjugacy separable* and
- *subgroup separable*;

a group G is subgroup separable if every finitely generated subgroup is closed in the profinite topology of G; this is equivalent to saying that for each finitely generated subgroup H, the property 'being in H' is a local-global property of elements of G. This has the important consequence that the closure of H in $\widehat{G}$ is naturally isomorphic to $\widehat{H}$.

That free groups are subgroup separable was proved by Marshall Hall in 1949 (see [LS], Chapter 1, Prop. 3.10). The fact that virtually polycyclic groups are subgroup separable is quite elementary (see [S], Chapter 1). The fact that they are conjugacy separable, however, depends on an interesting result in algebraic number theory (due to F. K. Schmidt and Chevalley):

Theorem 2.2.2. *Let $\mathcal{O}$ be the ring of integers in an algebraic number field, with group of units $\mathcal{O}^*$. Then every subgroup of finite index in $\mathcal{O}^*$ contains a 'congruence subgroup'* $(1 + m\mathcal{O}) \cap \mathcal{O}^*$ $(0 \neq m \in \mathbb{Z})$.

One says that $\mathcal{O}^*$ has the *congruence subgroup property*: equivalently, the profinite topology on the *additive* group of $\mathcal{O}$ induces the profinite topology on the *multiplicative* group $\mathcal{O}^*$. More generally, let M be any virtually polycyclic group and G a group of automorphisms of M. We may define a topology on G by choosing as a base for the neighbourhoods of 1 the family of subgroups

$$\mathrm{C}_G(M/M^m) \ (m \in \mathbb{N});$$

this is the *congruence topology* on G.

Theorem 2.2.3. *Let M be a virtually polycyclic group and G a virtually polycyclic subgroup of* $\mathrm{Aut}(M)$. *Then*

(i) *Each orbit a^G $(a \in M)$ is closed in the profinite topology of M;*
(ii) *G is closed in the congruence topology of* $\mathrm{Aut}(M)$;
(iii) *the congruence topology on G is the same as the profinite topology.*

Theorem 2.2.2 is the special case of (iii) where $M = \mathcal{O}$ and $G = \mathcal{O}^*$ (acting by multiplication). Part (iii) follows directly from (ii) applied to arbitrary subgroups of finite index in G. Part (ii) follows from (i) applied to the orbit of $(a_1, \dots, a_d)$ in the group $M^{(d)}$, where $\{a_1, \dots, a_d\}$ is a generating set for M and G acts diagonally. Part (i), a generalized version of conjugacy separability, may be reduced to an application of Theorem 2.2.2 by 'dévissage', arguing by induction on the Hirsch length of M. See [S], Chapter 4 (I assumed there that M is free abelian, but the general case is no harder). A further generalization is given in [S1], §8.

There is an essentially equivalent formulation of (i) in terms of *derivations*: a derivation from G to M is a map $\delta : G \to M$ such that $\delta(xy) = \delta(x)^y.\delta(y)$ for all $x, y \in G$ (such maps are also called *crossed homomorphisms* or 1-*cocyles*). Among these are the *inner derivations* $\delta_a : x \mapsto a^x a^{-1}$ $(a \in M$ fixed). Since $a^G = \delta_a(G) \cdot a$ we see that (i) is a special case of

Theorem 2.2.4. *Let M and G be virtually polycyclic groups, with G acting on M. If $\delta : G \to M$ is a derivation then the set $\delta(G)$ is closed in the profinite topology of M.*

The action of G on M induces an action of $\widehat{G}$ on $\widehat{M}$, and a derivation $\delta : G \to M$ induces a continuous derivation $\widehat{\delta} : \widehat{G} \to \widehat{M}$. One may deduce

Theorem 2.2.5. *Let G and M be as above. Then the natural mapping*

$$H^1(G, M) \to H^1(\widehat{G}, \widehat{M})$$

is injective.

Here, $H^1(G, M)$ is the 'non-abelian cohomology' set defined in [CG], Chapter 1. Another application of Theorem 2.2.4 gives

Proposition 2.2.6. *Let $a \in M$. Then $\widehat{\delta}^{-1}(a)$ is equal to the closure of $\delta^{-1}(a)$ in $\widehat{G}$.*

This applies in particular when δ is a homomorphism, and shows that the functor $G \mapsto \widehat{G}$ is *exact* on virtually polycyclic groups. This can also be seen by a direct elementary argument; but the following excellent properties of this functor depend on the full strength of Proposition 2.2.6:

Theorem 2.2.7. *Let G be a virtually polycyclic group and H, K subgroups of G. Then*

$$\begin{aligned} \mathrm{C}_{\bar{K}}(\bar{H}) &= \overline{\mathrm{C}_K(H)} \\ \mathrm{N}_{\bar{K}}(\bar{H}) &= \overline{\mathrm{N}_K(H)} \\ \bar{H} \cap \bar{K} &= \overline{H \cap K}. \end{aligned}$$

where $\bar{X}$ denotes the closure of a set X in $\widehat{G}$.

See [RSZ], §2. This is applied together with the geometric study of profinite groups acting on 'profinite trees' to establish

Theorem 2.2.8 ([RSZ]). *Let G a group that is obtained from virtually free groups and virtually polycyclic groups by forming finitely many successive free products, amalgamating cyclic subgroups. Then G is conjugacy separable.*

Sometimes, a purely global result can be deduced by a 'local-global' argument from the finite case: this is how Hirsch (in 1954) proved that the Frattini subgroup of a polycyclic group is nilpotent. As an application of Theorem 2.2.7, we show in [NS3] that if G is a virtually polycyclic group, N is a normal subgroup, and G is isomorphic to $N \times G/N$, then N is actually a direct factor of G; this is deduced from the special case of finite groups, a recent theorem due to J. Ayoub.

Another famous decision problem in group theory is the *isomorphism problem*: to decide, given two finite group presentations, whether or not they define isomorphic groups. Suppose $\mathcal{C}$ is a class of groups having the 'local-global property for isomorphism' – that is, for G and H in $\mathcal{C}$ one has $G \cong H$ if and only if $\mathcal{F}(G) = \mathcal{F}(H)$, where $\mathcal{F}(G)$ denotes the set of isomorphism types of finite quotient groups of G. Then it is easy to see, by a modification of the argument above, that the isomorphism problem for finitely presented groups in $\mathcal{C}$ has a positive solution. Examples of such classes $\mathcal{C}$ are the finitely generated free groups and the finitely generated abelian groups. As polycyclic groups are not so very different from finitely generated abelian groups, one might wonder whether they, also, have the local-global property for isomorphism. The answer is 'no': examples demonstrating this are given in [S], Chapter 11. Some of these examples are constructed using integral quadratic forms that are 'locally equivalent' but not equivalent over $\mathbb{Z}$. Such quadratic forms, however, do belong to the same genus, which consists of *finitely many* integral equivalence classes. And this finiteness property does indeed generalize:

Theorem 2.2.9 ([GPS]). *Given any set $\mathcal{X}$ of isomorphism types of finite groups, there are at most finitely many isomorphism types of virtually polycyclic groups G such that $\mathcal{F}(G) = \mathcal{X}$.*

The proof does not tell us exactly *how many* isomorphism types, so the theorem does *not* imply a positive solution for the isomorphism problem in this case. That requires other methods, and may be found in [S1]. While the statement of the theorem does not explicitly mention profinite groups, it is clear (if G is a finitely generated group!) that the set $\mathcal{F}(G)$ both determines and is determined by (the *topological* group) $\widehat{G}$, so the result amounts to saying that for virtually polycyclic groups, the profinite completion 'determines the group up to finitely many possibilities'. (In fact, in this case the topological group $\widehat{G}$ is uniquely determined by its underlying abstract group: see §2.6 below.)

Given any subgroup G of a profinite group P, the inclusion $G \to P$ induces a natural continuous homomorphism $\pi : \widehat{G} \to P$. This is surjective if and only if G is dense in P; it is injective if and only if the topology induced on G as a subspace of P is the profinite topology of G, in which case we say that G has the *congruence subgroup property*, or CSP (by analogy with the special case $P = \mathrm{Aut}(M)$ discussed above). Now we can reformulate Theorem 2.2.9 as

Theorem 2.2.10. *Let P be a profinite group. Let $\mathcal{S}$ denote the set of all virtually polycyclic subgroups that are dense in P and have CSP. Then $\mathcal{S}$ consists of finitely many orbits of* $\mathrm{Aut}(P)$.

In fact, using Theorem 2.4.1, below, one can show that if P is the profinite completion of a virtually polycyclic group, then any finitely generated residually finite group G with $\widehat{G} \cong P$ is itself virtually polycyclic (by considering the dimension of the Sylow pro-p subgroups of $\widehat{G}$: see §2.3). So in Theorem 2.2.10 we can replace 'virtually polycyclic' by 'finitely generated' as long as we add the hypothesis that P contain at least one dense virtually polycyclic subgroup with CSP.

The advantage (indeed, the necessity) of this 'profinite' approach is apparent as soon as one embarks on the proof of this theorem. One of the first steps, for example, is to show that for $G \in \mathcal{S}$ the closure in P of the Fitting subgroup of G is precisely the Fitting subgroup of P. Since G is subgroup separable it follows that $\widehat{\mathrm{Fit}(G)}$ is determined by $\widehat{G}$; the problem can now be broken into two cases: (1) the case of *nilpotent* groups, (2) the study of groups G for which not only $\widehat{G}$ but also $\mathrm{Fit}(G)$ are fixed. Both parts are difficult, and depend on deep results in the arithmetic theory of algebraic groups, results that generalize classical finiteness properties of quadratic forms. The key fact is the following analogue of Theorem 2.2.3(i):

Theorem 2.2.11 (Borel and Serre). *Let Γ be an arithmetic group acting rationally on $M = \mathbb{Z}^d$. Then each 'local orbit' of Γ in M is the union of finitely many orbits of Γ.*

A *local orbit* here means a set of the form $M \cap a^{\widetilde{\Gamma}}$ where $\widetilde{\Gamma}$ is the *integral adele group* associated to Γ, acting on $\widehat{M}$. For all this, see [S], Chapters 9 and 10.

A much simpler question than that of isomorphism is the following: what is the minimal size of a generating set for a group G? This number is denoted $\mathrm{d}(G)$.

Theorem 2.2.12 ([LW]). *If G is a virtually polycyclic group then* $\mathrm{d}(G) \leq \mathrm{d}(\widehat{G}) + 1$.

Of course, $\mathrm{d}(\widehat{G})$ here denotes the minimal size of a *topological* generating set for $\widehat{G}$, so what the result is saying is that if every finite quotient of G can be generated by d elements, then G itself can be generated by $d + 1$ elements. This is a hard theorem due to Linnell and Warhurst. An illuminating new proof has recently been obtained by Kassabov and Nikolov [KN2]. It is very easy to find cases where $\mathrm{d}(G) = \mathrm{d}(\widehat{G})$ (abelian groups for example), and not much harder to find cases where $\mathrm{d}(G) = \mathrm{d}(\widehat{G})+1$ (using a ring of algebraic integers that is not a PID). If we had an algorithm for deciding whether a polycyclic group is of the first or of the second type, we could then effectively determine $\mathrm{d}(G)$ for such groups G, by a version of the 'stupid double-enumeration procedure' described above. But – as far as I know – we don't. Indeed the following challenging problem is still open (even for the 'easy' case of virtually abelian groups!):

Problem. Find an algorithm that determines $\mathrm{d}(G)$ for every polycyclic group G.

For the currently known decision procedures for polycyclic groups see [BCRS], [S1], [E1] and [E2].

In contrast to Theorem 2.2.12, one can't in general infer anything about $\mathrm{d}(G)$ from knowledge of $\mathrm{d}(\widehat{G})$: an interesting construction by D. Wise produces finitely presented, residually finite, groups G with $\mathrm{d}(G)$ arbitrarily large and $\mathrm{d}(\widehat{G}) \leq 3$; see [Wi], Example 3.9.

A uniform bound for $\mathrm{d}(\bar{G})$ over all the finite images $\bar{G}$ of a group G is just one example of what I call an 'upper finiteness condition': a uniform bound for some measure of size, or growth, on all the finite quotients of a group. Any such condition certainly means something for the global structure of a group, and the challenge is to find out what it is. This programme is discussed in Section 2.4 below.

2.3. p-Adic analytic groups

The theory of Lie groups is without doubt one of the central pillars of twentieth-century mathematics (not to mention physics!). Quite early in the century, an analogous theory of 'p-adic Lie groups' received some attention: these *p-adic analytic groups* have the underlying structure of an analytic manifold over the field $\mathbb{Q}_p$, and the group operations are given locally by convergent p-adic power series. The global structure and cohomology theory of p-adic analytic groups were elucidated by Michel Lazard in a magisterial paper [L], published in 1965. One of his key discoveries was that each compact p-adic analytic group has an open subgroup (necessarily of finite index) which is a finitely generated pro-p group, and any pro-p group arising in this situation has a certain special algebraic property; conversely, every finite extension of a finitely generated pro-p group with this property has, in a natural way, the structure of a compact p-adic analytic group.

The 'special property' discovered by Lazard is that of being *powerful*, a term introduced later by Lubotzky and Mann in [LM]. The pro-p group G is powerful if G/G^p is abelian (when $p = 2$ we require that G/G^4 be abelian). Thus powerful groups are 'abelian to a first approximation', and Lubotzky and Mann went on to show that in fact such groups resemble abelian groups in many ways: for example, in a d-generator powerful group every closed subgroup can be generated by d elements. Thus such a group has *finite rank*, where the rank of a profinite group G is defined by

$$\mathrm{rk}(G) = \sup\{\mathrm{d}(H) | H \leq_c G\}$$

(here $d(H)$ denotes the minimal size of a (topological) generating set for H, and $H \leq_c G$ means 'H is a closed subgroup of G'). Conversely, they proved that every pro-p group of finite rank has an open (hence of finite index) powerful subgroup. With Lazard's result, this shows that *a pro-p group is p-adic analytic if and only if it has finite rank.*

This opened the way to a more group-theoretic approach to the whole topic, expounded in detail in the book [DDMS]. The resulting theory has found numerous applications. Applications to finite p-groups are discussed in §2.5 below. Many applications to infinite group theory are based on Lubotzky's observation that a compact p-adic analytic group is a linear group over $\mathbb{Q}_p$, by Ado's Theorem: it should be mentioned that the correspondence Lie groups $\leftrightarrow$ Lie algebras works even better in the p-adic case than in the classical case. This leads to the 'Lubotzky linearity criterion', see §2.6. It implies that any infinite group which is residually a finite p-group and whose pro-p completion has finite rank is in fact a linear group; such a group can then be attacked with various tools from linear group theory. A strikingly successful example of this strategy is discussed in the following section.

Other group-theoretic applications are described in [DDMS]. Our relatively good understanding of pro-p groups of finite rank has encouraged the investigation of wider classes of pro-p groups, and this is currently a lively area of research. Many recent developments are described in the book [NH].

2.4. Upper finiteness conditions and subgroup growth

2.4.1. *'Upper finiteness conditions'*

Let us consider the implications for a group of imposing various restrictions on its finite quotients.

1. The *rank* $\mathrm{rk}(Q)$ of a finite group Q is the least integer r such that every subgroup of Q can be generated by r elements. The *upper rank* of any group G is

$$\mathrm{ur}(G) = \sup\{\mathrm{rk}(Q) | Q \in \mathcal{F}(G)\}.$$

This is none other than the rank of $\widehat{G}$, defined above.

Theorem 2.4.1 ([MS1]). *Let G be a finitely generated residually finite group. Then G has finite upper rank if and only if G is virtually soluble of finite rank.*

An infinite (abstract) group G is said to have finite rank if there exists an integer r such that every *finitely generated* subgroup of G can be generated by r elements. Soluble groups of finite rank are quite easy to describe: such a group that is also finitely generated and residually finite is a finite extension of a triangular matrix group over a ring of the form $\mathbb{Z}[1/m]$. This theorem is making two remarkable assertions: (a) that '(bounded) finite rank' is a local-global property, and (b) that a *numerical bound* (on the size of generating sets, in this case) implies a *structural algebraic* property, namely solubility.

2. For a finite group Q, the *number of subgroups* of Q is denoted $s(Q)$. A group G has *weak polynomial subgroup growth*, or wPSG, if there exists a constant α such that

$$s(Q) \leq |Q|^{\alpha} \tag{2.4.1}$$

for every $Q \in \mathcal{F}(G)$.

Theorem 2.4.2 ([LMS], [S2]). *Let G be a finitely generated residually finite group. Then G has wPSG if and only if G is virtually soluble of finite rank.*

The alert reader will have noticed that this theorem implies the preceding one, since if G has finite upper rank we can take $\alpha = \mathrm{ur}(G)$ and deduce that G has wPSG.

3. A group G has *polynomial index growth*, or PIG, if there exists a constant α such that

$$|Q| \leq (\exp Q)^{\alpha}$$

for every $Q \in \mathcal{F}(G)$, where $\exp Q$ denotes the exponent of Q. This is equivalent to saying that $|Q/Q^m| \leq m^{\alpha}$ for every $Q \in \mathcal{F}(G)$ and every $m \in \mathbb{N}$. It is easy to see that every soluble group of finite rank has PIG, but the converse is far from true: Balog, Mann and Pyber [BMP] construct a finitely generated residually finite group with PIG which has finite simple quotients of unbounded ranks. However, if we *assume* solubility we have

Theorem 2.4.3 ([PS]). *Let G be a finitely generated soluble residually finite group. Then G has PIG if and only if G has finite rank.*

PIG and other upper finiteness conditions are discussed in detail in Chapter 12 of [SG] (where this last result appears as an open problem).

2.4.2. *Subgroup growth*

A group G has 'weak PSG' if it doesn't have very many subgroups of each finite index. More generally, it is interesting to study just how many subgroups there are of each index: that is, to study the function $n \mapsto a_n(G)$ where $a_n(G)$ denotes the number of subgroups of index n in G. This function is well defined as long as G is finitely generated. When G is a profinite group, $a_n(G)$ denotes the number of *open* subgroups of index n in G, and again is well defined if G is (topologically) finitely generated. Moreover, it is easy to see that if G is any abstract group, then $a_n(G) = a_n(\widehat{G})$; in this sense, *subgroup growth* – i.e. the behaviour of the function $n \mapsto a_n(G)$ – is a 'profinite' property of groups.

A comprehensive account of this topic is given in the book [SG], where the advantages of the 'profinite philosophy' are amply illustrated; let me just mention a few of the highlights, under three headings. We will denote by $s_n(G)$ the number of subgroups (or open subgroups) of index *at most* n in the group G.

Analytic problems. What does a given restriction on the subgroup growth imply for the algebraic structure of a group?

A group G has *polynomial subgroup growth*, or PSG, if $\log s_n(G) = O(\log n)$. This obviously implies wPSG, and it is a deep result (depending on CFSG*) that the two conditions are in fact equivalent. Thus the theorem stated above is equivalent to

Theorem 2.4.4 ([LMS]). *Let G be a finitely generated residually finite group. Then G has PSG if and only if G is virtually soluble of finite rank.*

The difficult part is 'only if'. The original proof of this (though not the one presented in [SG]) starts by considering the pro-p completions of G. Lubotzky and Mann proved that every pro-p group with PSG is p-adic analytic, from which it follows that if G has PSG then $\widehat{G}_p$ is a p-adic analytic group, and therefore *linear*. Thus if G happens to embed into $\widehat{G}_p$ then G itself is a linear group. One can then use 'Strong Approximation' results (specifically, Theorem 2.6.3 stated in §2.6, below) to reduce to the case of arithmetic groups, and the proof is concluded by an explicit counting of congruence subgroups in such groups (see §2.8.1). In the general case, further arguments are required, depending among other things on CFSG.

While a finitely generated residually finite group with PSG must be virtually soluble, this is not true for finitely generated *profinite* groups with PSG. These are characterized in [SSh]: such a profinite group is (virtually) an extension of a prosoluble group of finite rank by the Cartesian product of a family of finite quasisimple groups of Lie type satisfying certain very precise arithmetical conditions. (In view of the preceding theorem, such a group can only be the profinite completion of a finitely generated abstract group in the special case where this family of quasisimple groups is *finite*.)

Like much of 'pure' profinite group theory, the characterization of profinite groups with PSG quickly reduces to a problem of *finite* group theory: establishing uniform bounds for several structural parameters of a finite group G in terms of the parameter α defined in (2.4.1), above. The same applies to many other results that relate the algebraic structure of a profinite group to its rate of subgroup growth, when this is faster than polynomial.

Synthetic problems. Under this heading comes the problem of constructing groups that demonstrate particular types of subgroup growth. A group G is said to have *growth type f* if

$$\log s_n(G) = O(\log f(n))$$
$$\log s_n(G) \neq o(\log f(n)).$$

It is not difficult to construct finitely generated profinite groups with more-or-less arbitrary growth type, by forming Cartesian products of suitable collections of finite groups [MS2]. To do the same for finitely generated *abstract* groups is much harder, but we have

Theorem 2.4.5 ([P], [S4]). *Let $g : \mathbb{N} \to \mathbb{R}_+$ be a 'good' non-decreasing function with $g(n) = O(n)$. Then there exists a finitely generated group G having growth type $n^{g(n)}$.*

*The classification of the finite simple groups.

The condition 'good' here is a mild restriction of a technical nature, that need not concern us. The bound $g(n) = O(n)$ is necessary, because the fastest possible growth type for any finitely generated group is easily seen to be n^n. Thus the point of the theorem is that essentially every 'not impossible' growth type is actually exhibited by some finitely generated group.

The proof is in two stages. The first is to construct a suitable profinite group P with the specified growth type; the second, harder part, is to show that this P is the profinite completion of some finitely generated abstract group (this is what 'suitable' means here: the easy groups given in [MS2] don't have this property). That is, we require P to contain a dense finitely generated subgroup G that has the *congruence subgroup property*, as defined in §2.2, above. In fact two different constructions are used: when $g(n) = O(\log \log n)$ one takes P to be a certain group of automorphisms of a rooted tree; this construction is discussed in §2.6, below. When $\log n = O(g(n))$ one takes P to be the Cartesian product of a suitable family of finite alternating groups; in this case, the dense subgroup G does not quite have the CSP, but close enough: it turns out that the kernel of the natural epimorphism $\widehat{G} \to P$ is a procyclic group, which is enough to ensure that G has the same subgroup growth type as P. For full details see Chapter 13 of [SG] (a different and more general construction has recently been obtained in [KN]; see Theorem 2.6.4 below).

Zeta functions. Having associated to a finitely generated group G the numerical sequence $(a_n(G))$, it is natural to wonder about the arithmetical properties of this sequence. The 'growth type' defined above is one crude measure, but can we obtain more refined information? This question has been studied in depth for certain types of groups: (a) free groups, one-relator groups and free products of finite groups, (b) finitely generated nilpotent groups, and (c) p-adic analytic pro-p groups.

I will say no more about the class (a). This is the subject of many papers by Thomas Müller, using methods of combinatorics and analysis; for references and some sample results see Chapter 14 of [SG]. Groups of types (b) and (c) have polynomial subgroup growth: in this case, it is convenient to encode the sequence $a_n(G)$ in a generating function

$$\zeta_G(s) = \sum_{n=1}^{\infty} a_n(G) n^{-s}$$

where s is a complex variable. This 'zeta function' represents a complex analytic function, regular on some half-plane $Re(s) > \alpha$; here the abscissa of convergence α is given by

$$\alpha = \inf\{\gamma \,|s_n(G) = O(n^{\gamma})\},$$

a finite number when G has PSG.

For a fixed nilpotent group G, it is easy to see that the arithmetical function $a_n(G)$ is multiplicative, i.e. if m and n are coprime then $a_{mn}(G) = a_m(G)a_n(G)$. This implies the 'Euler product' decomposition

$$\zeta_G(s) = \prod_p \zeta_{G,p}(s)$$

where the product is over all primes and the 'local factors' are defined by

$$\zeta_{G,p}(s) = \sum_{j=0}^{\infty} a_{p^j}(G) p^{-js}.$$

We showed in [GSS] that when G is a finitely generated nilpotent group, for each prime p the series $\zeta_{G,p}(s)$ represents a *rational function* in p^{-s} (with rational coefficients); to see why this is reasonable, note that when G is the infinite cyclic group ζ_G is the Riemann zeta function, and $\zeta_{G,p}(s) = \frac{1}{1-p^{-s}}$. The proof applies a general theorem about p-adic integrals, proved by Denef using methods of p-adic model theory. Now, still assuming that G is finitely generated and nilpotent, we have $\zeta_{G,p}(s) = \zeta_P(s)$ where $P = \widehat{G}_p$ is the pro-p completion of G; and P in this case is a p-adic analytic pro-p group. Thus the rationality theorem just mentioned is a very special case of

Theorem 2.4.6 ([dS1]). *If P is a compact p-adic analytic group then $\zeta_{G,p}(s)$ is a rational function over $\mathbb{Q}$ in p^{-s}.*

In order to establish this, du Sautoy showed that the 'analytic' theory of p-adic analytic groups can be reduced to 'p-adic analytic' model theory, as developed by Denef and van den Dries. As well as opening up a fascinating new field of study, this result led the way to some remarkable applications in the theory of finite p-groups, discussed in the following section.

The study of these group-theoretic zeta functions is a very active area of research at the present time; many results have been obtained but many more challenging problems remain open. For more details and references up to 2002 see [dSS] and [SG], Chapters 15 and 16 (but there has been much progress since then, for example in the work of Christopher Voll [V]). The recent book [dSW] presents a large range of explicit examples as well as some new theory and more up-to-date references.

Instead of counting subgroups of finite index, one could count equivalence classes of finite-dimensional representations; the Dirichlet series encoding these numbers give rise to 'representation-growth zeta functions'. See [J-Z1], [LL] and [V] for some recent and deep results about these.

2.5. Finite p-groups

2.5.1. *Coclass*

It was clear from the early days of group theory that the finite simple groups are rather special: they are, essentially, the symmetry groups of highly symmetrical structures (a finite set, or a vector space with a bilinear form). Of course this wasn't actually proved until the 1980s (and the final steps have only just been published), but the fact is that these objects form an elegant list of identifiable objects, and they are 'rigid' in two senses: (1) they are *isolated*: you can't move from one to the next by a small deformation, and (2) the possibilities of building composite groups out of them are very limited: they have small Schur multipliers and small outer automorphism groups.

Neither of these (slightly vague) statements is true of nilpotent groups. It was equally clear, at least from the 1930s with the work of Philip Hall and others, that the finite p-groups constitute a vast and rather amorphous collection. Thus the received wisdom for most of the last century considered finite p-groups to be unclassifiable.

This pessimistic conclusion was based on the experience of trying to produce coherent lists of p-groups, starting with the smallest and working up by size; in practice this was only achieved for groups of nilpotency class 2 and quite modest size, as the number of groups of order p^n was found to grow extremely fast with n. Higman and Sims showed in the 1960s that this number is about $p^{\frac{2}{27}n^3}$, and that the number of groups of class 2 is already about this big. (Contrast this with the number of simple groups of order n, which is nearly always zero, sometimes one and very occasionally two!)

However, a different picture appears if instead of small nilpotency class one looks at p-groups of *large* class. Completing earlier work of Blackburn, Leedham-Green and McKay found that the *p-groups of maximal class* do form a comprehensible pattern, and can indeed be neatly classified by their order. What emerged from this classification is that, for a fixed prime p, the best way to think of p-groups of maximal class is as the finite quotients of one particular pro-p group; for example, the 2-groups of maximal class are precisely the finite quotients of the 'dihedral pro-2 group' $\mathbb{Z}_2 \rtimes C_2$, together with certain natural 'twistings' of them (quaternion or semi-dihedral groups). This realisation led Leedham-Green and Newman to formulate an audacious generalization, that became known as the "coclass conjectures". These profoundly insightful conjectures cast the problem of classifying p-groups into a completely new framework, and totally transformed the subject between 1980 and 1994, when the conjectures were finally established.

A finite p-group is said to have *coclass* r if it has order p^n and nilpotency class $n-r$ (so maximal class means coclass 1). A pro-p group has coclass r if it is the inverse limit of a system of finite p -groups of coclass r (with all maps surjective). The main conjecture of Leedham-Green and Newman, Conjecture A, is purely finitary: it states that every p-group of coclass r has a normal subgroup of nilpotency class at most 2 and bounded index (the bound depending only on p and r). In view of the remarks above, this might seem like no progress as regards the classification: what lies behind it, however, is a vision of the whole universe of p-groups of fixed coclass. For given p and r, one arranges the set of all (isomorphism types) of coclass r p-groups into a graph $\mathcal{G}(p,r)$, whose directed edges represent the quotient maps $G \to G/Z$ where Z is a central subgroup of order p in G. Each infinite chain in this graph then gives rise in a natural way to a pro-p group of coclass r. Now the remarkable facts are these:

- There are only finitely many infinite pro-p groups of coclass r (for given p and r);
- Each infinite pro-p group of finite coclass is finitely generated and virtually abelian, in other words, it is a finite extension of $\mathbb{Z}_p^d$ for some finite d.

Moreover, every finite p-group of coclass r is either a quotient of one of these virtually abelian pro-p groups, or is obtained from such a quotient by an explicit 'twisting' process, or is one of finitely many 'sporadic' groups.

A key step in the proof, achieved by Leedham-Green, was to show that every pro-p group of finite coclass is a *p-adic analytic group*, that is, a pro-p group of finite rank. Once this was known, it became possible to apply powerful techniques for studying such groups, to show that if a p-adic analytic group has finite coclass then it must be virtually abelian. The first proof of this fact, due to Donkin, rests on the 'analytic' aspect of these groups and applies the classification of semisimple p-adic Lie algebras, thus establishing a bridge between the theory of p-groups and the theory of finite simple groups. Subsequently, a clever direct argument (also using Lie algebras) was found by Shalev and Zelmanov. An alternative, purely finitary, proof for Conjecture A was later obtained by Shalev [Sh]; although this avoids the use of pro-p groups altogether, it was clearly inspired by the p-adic methods used before.

Explanatory accounts of all or parts of this story are to be found in [LGM1], [LGM2], [DDMS], Chapter 10. For full references to the many original papers, see the bibliographies to [LGM1] and [LGM2].

2.5.2. *Conjecture P*

The main results of coclass theory show that the graph $\mathcal{G}(p, r)$ has finitely many components; moreover, if we remove a finite number of 'sporadic' groups what remains is the disjoint union of finitely many trees. Each of these trees contains just one maximal infinite chain, the 'trunk', to which are attached infinitely many finite 'twigs'. On the basis of extensive computer investigations, M. Newman and E. O'Brien were led to make some very precise conjectures about the shape of these trees. In particular, their *Conjecture P* asserts that when $p = 2$, each tree is eventually periodic, with period dividing 2^{r-1}.

The conjecture obviously implies that the twigs of such a tree are of bounded length, and this is no longer true when the prime p is odd. However, du Sautoy was able to establish a general periodicity result which includes (the qualitative part of) Conjecture P as a special case. For each tree $\mathcal{T}$ as above and each natural number m, let $\mathcal{T}[m]$ denote the 'pruned tree' obtained from $\mathcal{T}$ by removing all vertices whose distance from the trunk exceeds m.

Theorem 2.5.1 ([dS2]). *Each of the pruned trees $\mathcal{T}[m]$ is eventually periodic.*

It is known that when $p = 2$ the twigs have bounded lengths, so in this case we have $\mathcal{T}[m] = \mathcal{T}$ for some value of m.

The proof is a remarkable application of du Sautoy's rationality theorem for zeta functions (see §2.4 above). First of all, he deduces from the results of coclass theory that there exists a certain p-adic analytic pro-p group $H = H(p, r)$ which maps onto every finite p-group of coclass r. The holomorph $P = H \rtimes \mathrm{Aut}(H)$ is again a p-adic analytic group, and du Sautoy associates a certain generalized zeta function to the pair (H, P); the coefficients of (the Dirichlet series defining) this function encode precisely the 'shape' of the pruned tree $\mathcal{T}[m]$. He proves that this generalized zeta function is again rational, and the stated periodicity then emerges as a formal consequence. For details of this argument, see [dSS]. A group-theoretic proof, which yields more detailed information, has recently been obtained by Eick and Leedham–Green [EL].

Zeta functions are also used in [dS2] to obtain results about the enumeration of p-groups and of finite nilpotent groups of fixed *nilpotency class*. These are also discussed in [dSS].

2.6. Finitely generated groups

2.6.1. *Linearity*

Nearly a century ago, Hasse argued in favour of treating the p-adic completions of $\mathbb{Q}$ on the same footing as the reals. This idea had a huge influence on the development of number theory; as mentioned above, it also led to the idea of studying (non-commutative) groups via their profinite completions. In general, a group deosn't even have a 'real completion' (unless it is nilpotent, say), but every group has its pro-p completions and its profinite completion. Thus every group can be mapped, functorially, into various interesting compact topological groups.

This simple idea led Lubotzky to the solution of a long-standing problem: how to characterize, by purely internal criteria, those groups that have a faithful finite-dimensional linear representation over some field, 'linear groups' for short.

Theorem 2.6.1 ([Lu2] 'Lubotzky linearity criterion'). *Let G be a finitely generated group. Then G is linear over some field of characteristic zero if and only if, for some prime p and some natural number r, G has a chain of normal subgroups*

$$G = G_0 \geq G_1 \geq G_2 \geq \dots$$

such that (i) *G/G_1 is finite,* (ii) *G_1/G_n is a finite p-group of rank at most r for every $n \geq 1$, and* (iii) $\bigcap_n G_n = 1$.

Suppose that G satisfies the given condition, and consider the inverse limit

$$P = \varprojlim G_1/G_n.$$

Hypothesis (ii) implies that this group P is a pro-p group of finite rank, and so a p-adic analytic group (see §2.3 above). Then Lie theory and Ado's Theorem show that P is linear over the p-adic number field $\mathbb{Q}_p$. Hypothesis (iii) implies that G_1 embeds into P, so G_1 is linear, and it follows by hypothesis (i) that G itself is linear (form the induced representation).

Note that the argument so far does not require G to be finitely generated; the converse, however does. To see why it is true, suppose now that G is a finitely generated subgroup of $\mathrm{GL}_d(F)$ where F is a field of characteristic zero. Then in fact $G \leq \mathrm{GL}_d(R)$ where R is some finitely generated subring of F. Commutative algebra shows that for almost all primes p, such a ring R can be embedded in a matrix ring over $\mathbb{Z}_p$; for each such prime it follows that G can be embedded in some $\mathrm{GL}_{d'}(\mathbb{Z}_p)$

(where $d' = md$ may depend on p). Choosing a suitable prime p and identifying G with its image in $\mathrm{GL}_{d'}(\mathbb{Z}_p)$, we take

$$G_n = \{g \in G | g \equiv \mathbf{1} \,(\mathrm{mod}\; p^n)\}.$$

It is easy to see that the sequence (G_n) then satisfies conditions (i) and (iii); and condition (ii) is satisfied because the 'first congruence subgroup'

$$\mathrm{GL}^1_{d'}(\mathbb{Z}_p) = \ker(\mathrm{GL}_{d'}(\mathbb{Z}_p) \to \mathrm{GL}_{d'}(\mathbb{Z}_p/p\mathbb{Z}_p))$$

is a pro-p group of finite rank ([DDMS], Chapter 5).

For a more detailed account, and several variations on the same theme, see [DDMS] Interlude B.

So far, no-one has succeeded in establishing, or even formulating, an analogous characterization of the finitely generated linear groups over fields of positive characteristic, and this remains a challenging open problem. Lubotzky's criterion can paraphrased as: "some pro-p completion of some normal subgroup of finite index in G is p-adic analytic"; a natural starting point for the characteristic-p analogue would be to gain a better understanding of the pro-p groups that are 'analytic' over a local ring of characteristic p; the beginnings of such a theory are outlined in the final chapter of [DDMS].

The 'Lubotzky criterion' arises from considering congruence subgroups modulo powers of a fixed prime – looking 'downwards', we may say. Another way of looking at a finitely generated linear group is 'sideways': for example, we can embed $\mathrm{GL}_d(\mathbb{Z})$ into the Cartesian product $\prod_p \mathrm{GL}_d(\mathbb{Z}/p\mathbb{Z})$, where p ranges over any infinite set of primes. A. I. Mal'cev generalized this observation to show that every finitely generated linear group of degree d is residually 'linear of degree d over a finite field'. The precise converse is not true, but J. S. Wilson showed that a slightly weaker statement does hold: *if a finitely generated group G is residually (linear of degree d) then G is a subdirect product of finitely many linear groups.* For the proof, and some refinements, see [SG], Window 8. Here is one such refinement, which serves as a reduction step for many of the results stated in §2.4, above:

Theorem 2.6.2. *Let G be a finitely generated group and (N_i) a family of normal subgroups of G with $\bigcap_i N_i = 1$. Suppose that $G/N_i \leq \mathrm{GL}_d(F_i)$, where each F_i is either a field of characteristic zero or a finite field, and suppose further that for each prime p the number of i with* $\mathrm{char} F_i = p$ *is finite. Then G is linear over a field of characteristic zero.*

2.6.2. *Finite quotients*

What does it mean for a family of finite groups $\mathcal{X}$ to be precisely the set $\mathcal{F}(\Gamma)$ of (isomorphism types of) all finite quotients of some finitely generated group Γ? Equivalently, what does it mean for a profinite group G to be the profinite completion of a finitely generated (abstract) group? As mentioned in §2.2, this holds if and only if G contains a dense finitely generated subgroup Γ that has the congruence subgroup property; so the question may be seen as finding necessary and/or sufficient conditions on a profinite group G, expressed in terms of the family $\mathcal{F}(G)$, for the existence of such a subgroup (when G is a *profinite* group, $\mathcal{F}(G)$ denotes the set of *continuous* finite quotient groups of G: see the next subsection).

Two obvious necessary conditions for such a family $\mathcal{X}$ are (1) that $\mathcal{X}$ is quotient-closed, and (2) that all the groups in $\mathcal{X}$ can be generated by some bounded number of elements; but it seems very difficult to find further, less obvious ones. Suppose for example that $\mathcal{X}$ contains a subgroup X_i of $\mathrm{GL}_d(F_i)$ for $i = 1, 2, \ldots$ where F_i is a finite field of characteristic p_i and $p_1, p_2, \ldots$ is an infinite sequence of distinct primes. Then Γ has a quotient $\bar{\Gamma}$ which satisfies the hypotheses of Theorem 2.6.2, so $\bar{\Gamma}$ is a finitely generated characteristic-zero linear group; if we assume also that the groups X_i are simple and of unbounded orders (or some suitable weaker condition), we find that $\bar{\Gamma}$ is not virtually soluble. Under these conditions, $\bar{\Gamma}$ is guaranteed to possess a host of special finite quotients: applying a deep 'strong approximation' theorem due to Nori and Weisfeiler, Lubotzky established the following important result:

Theorem 2.6.3 ('Lubotzky alternative'). *Let Γ be a finitely generated linear group over a field of characteristic zero. Then one of the following holds:*

(a) *Γ is virtually soluble;*
(b) *there exist a connected, simply connected simple algebraic group $\mathfrak{G}$ over $\mathbb{Q}$, a finite set of primes S such that $\mathfrak{G}(\mathbb{Z}_S)$ is infinite, and a subgroup Γ_1 of finite index in Γ such that the profinite group $\mathfrak{G}(\widehat{\mathbb{Z}_S})$ is an image of $\widehat{\Gamma}_1$.*

(Here $\mathbb{Z}_S = \mathbb{Z}\big[\frac{1}{p};\ p \in S\big]$, and $\mathfrak{G}(\widehat{\mathbb{Z}_S})$ is isomorphic to the product $\prod_{p\notin S} \mathfrak{G}(\mathbb{Z}_p)$.) For the proof, see [SG], Window 9. Applying this to the group $\bar{\Gamma}$, we may deduce that the set $\mathcal{X}$ must contain many other groups in addition to the X_i: for each prime $p \notin S$ and each n, a group Q_p containing $\mathfrak{G}(\mathbb{Z}/p^n\mathbb{Z})$ as a subgroup, the indices $|Q_p : \mathfrak{G}(\mathbb{Z}/p^n\mathbb{Z})|$ being bounded above by a constant.

Thus if $\mathcal{P}$ is an infinite set of primes, a set of groups like

$$\left\{ \prod_{p\in T} \mathrm{PSL}_d(\mathbb{F}_p) | T \text{ a finite subset of } \mathcal{P} \right\}$$

cannot be the whole of $\mathcal{F}(\Gamma)$ for a finitely generated group Γ, while of course it is equal to $\mathcal{F}(G)$ where $G = \prod_{p\in\mathcal{P}} \mathrm{PSL}_d(\mathbb{F}_p)$. Thus the 2-generator profinite group G cannot be the profinite completion of a finitely generated group.

The problem with this group G is that the finite simple factor groups have bounded ranks. In an amazing feat of ingenuity, Kassabov and Nikolov have recently shown that this is essentially the *only* obstacle, when it comes to products of finite simple groups. For a group S they write $l(S)$ to denote the largest integer k such that S contains a copy of the alternating group $\mathrm{Alt}(k)$, and they prove

Theorem 2.6.4 ([KN]). *Let (S_n) be a sequence of finite simple groups such that $l(S_n) \to \infty$ as $n \to \infty$, and let*

$$G = \prod_{n=1}^{\infty} S_n.$$

If G is finitely generated (as a profinite group), then G is the profinite completion of a finitely generated group.

Thus we now have some understanding of what it means for a family of *semisimple* groups (products of finite simple groups) to be equal to $\mathcal{F}(\Gamma)$ for some finitely generated group Γ. However, it seems difficult even to formulate a conjecture about the nature of sets like $\mathcal{F}(\Gamma)$ in general.

Meanwhile, we could consider weakening the question a little, and asking: what does it mean for a collection of finite simple groups to be precisely the collection of *composition factors* of groups in $\mathcal{F}(\Gamma)$ for some finitely generated group Γ? These are called the *upper composition factors* of Γ. An almost complete answer is provided by

Theorem 2.6.5 ([S4]). *Let $\mathcal{S}$ be any collection of (isomorphism types) of non-abelian finite simple groups. Then there exists a 63-generator group Γ whose set of upper composition factors is precisely $\mathcal{S}$.*

To construct such a group Γ we start with a suitable profinite group G, and then find Γ as a dense subgroup in G. To begin with, we enumerate $\mathcal{S}$ as $\{X_1, X_2, \ldots, X_n, \ldots\}$ (if $\mathcal{S}$ is finite, the result is trivial, given that every finite simple group can be generated by 2 elements: this follows from CFSG, and implies that $\Gamma = \prod_{X \in \mathcal{S}} X$ is a 2-generator group). For each n we pick a faithful primitive permutation representation for X_n, and so identify X_n with a subgroup of $\mathrm{Sym}(l_n)$ for some l_n. Take $W_1 = X_1$, for $n > 1$ let W_n be the permutational wreath product

$$W_n = X_n \wr W_{n-1},$$

and define

$$G = \varprojlim W_n.$$

Thus G is a profinite group, whose set of upper composition factors is precisely $\mathcal{S}$.

This is all very easy. The challenge now is to find a suitable dense subgroup in G. The key lies in realizing G as a group of automorphisms of a suitable object.

Given the sequence of positive integers (l_n), consider the *spherically homogeneous rooted tree $\mathcal{T}$* of type (l_n): this is a connected graph without circuits, having a distinguished vertex v_0 (the *root*), and for each $n \geq 1$ having $l_1 \ldots l_n$ vertices at distance n from the root, each of valency $1 + l_{n+1}$ (so at each vertex of 'level' $n \geq 1$ there is one edge pointing 'upwards' towards the root and l_{n+1} edges pointing 'downwards' to the next level). It is easy to see that the automorphism group of this structure is the inverse limit of the finite permutational wreath products

$$V_n = \mathrm{Sym}(l_n) \wr \ldots \wr \mathrm{Sym}(l_2) \wr \mathrm{Sym}(l_1).$$

Thus V_n contains W_n as a permutation group for each n, and we may identify our profinite group G as a closed subgroup of $\mathrm{Aut}(\mathcal{T})$; a base for the neighbourhoods of the identity in G is given by the 'level-stabilizers' $\mathrm{st}_G(n) = \ker(G \to W_n)$.

One of the main results of [S4] states that there exists a 61-generator perfect group P that maps onto every non-abelian finite simple group. Using this, we define 63 specific tree automorphisms of $\mathcal{T}$, all lying in the group G, and take Γ to be the group generated by these 63 automorphisms. These generators are so chosen that (a) for each n, the group Γ acts as the whole group W_n on the nth level of $\mathcal{T}$, and (b) each nontrivial normal subgroup of Γ contains $\mathrm{st}_\Gamma(n) = \Gamma \cap \mathrm{st}_G(n)$ for some n (actually, a quite general argument shows that each nontrivial normal subgroup of Γ contains the derived group $\mathrm{st}_\Gamma(n)'$ of $\mathrm{st}_\Gamma(n)$ for some n; the role of the perfect group P is to ensure that in our case we have $\mathrm{st}_\Gamma(n)' = \mathrm{st}_\Gamma(n)$ for each n). Property (a) means that Γ is dense in G, while property (b) implies that Γ has the CSP in G. It follows that $\mathcal{F}(\Gamma) = \mathcal{F}(G)$, and hence that the set of upper composition factors of Γ is precisely $\mathcal{S}$.

The same construction, using sets $\mathcal{S}$ of the form $\{\mathrm{PSL}_2(\mathbb{F}_p) | p \in \mathcal{P}\}$ for suitably chosen sets of primes $\mathcal{P}$, was used in [S4] to construct finitely generated groups with arbitrarily specified types of subgroup growth (within a certain range). For details, and more discussion of trees like $\mathcal{T}$, see Chapter 13 of [SG].

Certain groups of rooted tree automorphisms called *branch groups* have been studied in depth by Grigorchuk and others. These include the groups described above, but are more usually pro-p groups (or dense finitely generated subgroups thereof); the celebrated construction by Grigorchuk of a finitely generated group having 'intermediate word growth' was (a dense subgroup of) a pro-2 branch group. See [G] and [BG].

2.6.3. *Forgetting the topology*

To be given a profinite group G is more or less equivalent to being given the family of all finite continuous quotient groups of G, that is, the groups G/N where N ranges over all the open normal subgroups of G. Indeed, G is (naturally isomorphic to) the inverse limit of this family, relative to the natural quotient maps $G/N \to G/M$, $(M \geq N)$. If we forget the topology and think of G just as an abstract group, we would expect to lose a lot of information: out of all the normal subgroups of finite index in G, how could we possibly pick out those that were open? Consider the following simple example. Fix a prime p, for each i let C_i be cyclic of order p, put $G_n = C_1 \times \cdots \times C_n$ and let

$$G = \varprojlim G_n$$

where $G_m \to G_n$ for $m \geq n$ are the obvious projection maps. The open subgroups of G are those that contain $\ker(G \to G_n)$ for some n, so there are only countably many of them. On the other hand, as an abstract group G is abelian, of exponent p and uncountable (of cardinality $\mathfrak{c} = 2^{\aleph_0}$); it is therefore a $\mathfrak{c}$-dimensional vector space over $\mathbb{F}_p$ and so contains $2^{\mathfrak{c}}$ subspaces of finite codimension. Thus G has $2^{\mathfrak{c}}$ (normal) subgroups of finite index, of which only countably many are open. It is obvious, from the very homogeneous nature of (the abstract group) G, that there is no way of recovering the original topology. (A similar construction can be made using any nontrivial finite group in place of the group of order p: see [RZ2], Ex. 4.2.13.)

However: if we restrict attention to (topologically) *finitely generated* profinite groups, the opposite is true:

Theorem 2.6.6 ([NS2]). *In a finitely generated profinite group, every subgroup of finite index is open.*

This is a remarkable fact: if we form the inverse limit G of any (surjective) inverse system $\mathcal{S}$ of finite groups, all of which can be generated by some fixed number of elements, then the only finite groups onto which G can be mapped homomorphically are the quotients of groups in $\mathcal{S}$; moreover, since the subgroups of finite index form a base for the neighbourhoods of the identity, the topology of G is completely determined by its structure as an abstract group.

This theorem is a case where a problem on profinite groups served as the motivation for some new developments in finite group theory, and it illustrates very clearly the principle that a qualitative property of profinite groups corresponds to a *uniform* quantitative property of finite groups. The basic idea is as follows. Let $w = w(x_1, \ldots, x_k)$ be a group word, and G a profinite group. Since the mappings

$$(g_1, \ldots, g_k) \mapsto w(g_1, \ldots, g_k),$$
$$(g_1, \ldots, g_k) \mapsto w(g_1, \ldots, g_k)^{-1}$$

from $G^{(k)}$ to G are continuous, their images in G are compact. It follows that for each n, the set $S(n)$ of all products of n elements of the form $w(g_1, \ldots, g_k)^{\pm 1}$ is compact, hence closed in G. Now consider the *verbal subgroup* $w(G)$, generated (*algebraically*, not topologically) by all values of w in G:

$$w(G) = \bigcup_{n=1}^{\infty} S(n). \tag{2.6.1}$$

If it happens that for some finite n we have $w(G) = S(n)$, then $w(G)$ is closed; conversely, if $w(G)$ is closed then a simple argument using the Baire category theorem and (2.6.1) shows that $w(G) = S(n)$ for some n. This means that every product of w-values in G is equal to a product of n w-values (where by 'w-value' I mean an element of the form $w(g_1, \ldots, g_k)^{\pm 1}$); let me abbreviate this to 'w has width n in G'.

On the other hand, it is easy to see that w has width n in G if and only if w has width n in G/N for every open normal subgroup N of G. Indeed, if the latter holds then

$$\frac{w(G)N}{N} = w(G/N) = \frac{S(n)N}{N}$$

for each N, so

$$S(n) \subseteq w(G) \subseteq \bigcap_N w(G)N = \bigcap_N S(n)N = S(n), \tag{2.6.2}$$

the last equality holding because $S(n)$ is a closed subset of G. The converse is obvious. Thus we have established the link between a qualitative property of G and a uniform property of $\mathcal{F}(G)$ (the set of *continuous* finite images of G):

Proposition 2.6.7. *Let G be a profinite group and w a group word. Then the (algebraic) verbal subgroup $w(G)$ is closed in G if and only if there exists a natural number n such that w has width n in every continuous finite image of G.*

This result, due to Brian Hartley, is nice, but how does it help with our original problem? Suppose we know in addition that the index $|Q : w(Q)|$ is uniformly bounded for all $Q \in \mathcal{F}(G)$. Then the big intersections in the middle of (2.6.2) contain only finitely many distinct terms, each of which is an open subgroup of G; and we may infer that in this case, $w(G)$ is not only closed but *open*.

Now let G be a d-generator profinite group and H a subgroup of finite index. Then H contains a subgroup K which is normal and of finite index in G. Let $F = F_d$ be the free group on free generators $x_1, \ldots, x_d$ and let D be the intersection of the kernels of all homomorphisms $F \to G/K$. Then D has finite index in F and is therefore finitely generated, by $w_1(x_1, \ldots, x_d), \ldots, w_m(x_1, \ldots, x_d)$ say. Put

$$w(\mathbf{y}_1, \ldots, \mathbf{y}_m) = w_1(\mathbf{y}_1) \ldots w_m(\mathbf{y}_m)$$

where $\mathbf{y}_1, \ldots, \mathbf{y}_m$ are disjoint d-tuples of variables. It is easy to see that (i) $w(F) = D$ and (ii) $w(G) \leq K$. The latter shows that H will be open in G if $w(G)$ is open. Property (i) implies that

$$|Q : w(Q)| \leq |F : w(F)| < \infty$$

for every $Q \in \mathcal{F}(G)$ (while we know nothing, a priori, about the finite group G/K, we do know that each of the finite groups in $\mathcal{F}(G)$ is d-generator, hence an image of F). To conclude that $w(G)$, and therefore also H, is open in G, we are thus reduced to establishing the following 'uniformity theorem' about finite groups (I call w 'd-locally finite' if $|F_d : w(F_d)|$ is finite):

Theorem 2.6.8 ([NS2]). *Let d be a natural number and let w be a d-locally finite group word. Then there exists $f = f(w, d)$ such that w has width f in every d-generator finite group.*

The proof of this result is long and difficult, and depends on CFSG. I will say no more about it here; for a brief outline see the announcement [NS1].

In the same paper we establish an analogous theorem for the commutator words $w = [x_1, \ldots, x_k]$; in view of Proposition 2.6.7 this implies that the derived group, and the higher terms of the lower central series, are closed in every finitely generated profinite group. We also made a not entirely successful attempt to do the same for the words $w = x^q$ ($q \in \mathbb{N}$), so the following is still open:*

Problem. Let q be a natural number. Is it true that the subgroup $G^q = \langle g^q | g \in G \rangle$ (generated *algebraically* by all qth powers in G) is open in G, for every (topologically) finitely generated profinite group G?

Note that in this situation, G^q is open if and only if it is closed, because there is a finite upper bound for the order of every finite d-generator group of exponent dividing q: this is the positive solution of the restricted Burnside Problem, due to Zelmanov. Thus the problem is equivalent to asking whether, for each d, the "Burnside word" x^q has bounded width in all d-generator finite groups.

---

* *Added in proof* No longer; see N. Nikolov and D. Segal, Powers in finite groups, *to appear*.

Whatever the answer turns out to be, results of this type certainly don't hold for arbitrary words: Romankov [R] has given a simple construction for a three-generator soluble pro-p group G in which the second derived group G'' is not closed; and $G'' = w(G)$ where $w = [[x_1, x_2], [x_3, x_4]]$. It would be very interesting to find a characterization of those group words w which have the uniformity property of Theorem 2.6.8. This is equivalent to asking: *for which words w is it the case that $w(G)$ is closed in G for every finitely generated profinite group G ?*; if we restrict to pro-p groups, the remarkably simple answer has recently been discovered by Andrei Jaikin [J-Z2]: $w(G)$ is closed in G for every finitely generated pro-p group G if and only if $w \notin F''(F')^p$, where F is the free group on the variables occurring in w. This and other questions of 'verbal width' are comprehensively explored in the book [W].

Let us turn briefly to the non-finitely generated case. For a profinite group G, let G_0 denote the underlying abstract group. Theorem 2.6.6 implies that if G is finitely generated then $\mathcal{F}(G) = \mathcal{F}(G_0)$ (recall that these denote the sets of *isomorphism types* of finite quotients, by open normal subgroups or by all normal subgroups of finite index, respectively). We have also seen examples of (infinitely generated) profinite groups G that have many non-open normal subgroups of finite index; but in these examples, too, we have $\mathcal{F}(G) = \mathcal{F}(G_0)$ – the same finite groups appear, though with different multiplicities as quotients of G. To construct a group G such that $\mathcal{F}(G) \neq \mathcal{F}(G_0)$ takes a little more effort; the following example was suggested by Lubotzky and Holt. For a finite group $S = S^2$, let $f(S)$ denote the least integer n such that every element of S is equal to a product of n squares (here S^2 denotes the subgroup generated by all squares). Now for each n let S_n be a finite group with $S_n = S_n^2$ and $f(S_n) > n$, and take $G = \prod_{n=1}^{\infty} S_n$. Proposition 2.6.7 shows that the subgroup G^2 is not closed in G; in particular it can't be equal to G, so G/G^2 has the cyclic group C_2 of order 2 as a quotient. On the other hand, $C_2 \notin \mathcal{F}(G)$ since every continuous finite quotient of G is a quotient of $S_1 \times \cdots \times S_k$ for some k. Thus $C_2 \in \mathcal{F}(G_0) \backslash \mathcal{F}(G)$.

Suitable groups S_n may be constructed as follows (for details, see [H]). Let $H = \mathrm{SL}_2(\mathbb{F}_4)$ and let M be its natural 2-dimensional $\mathbb{F}_4$-module, considered as a 4-dimensional $\mathbb{F}_2 H$-module. There is an H-epimorphism ϕ from $M \otimes_{\mathbb{F}_2} M$ onto the trivial module $\mathbb{F}_2$. Now let $M_1, \ldots, M_k$ be copies of M and form a special 2-group P with $P/[P, P] = M_1 \times \cdots \times M_k$ and $\mathrm{Z}(P) = [P, P] = \prod_{i<j}[M_i, M_j]$, where $[M_i, M_j] \cong \mathbb{F}_2$ and the commutator mapping $M_i \times M_j \to [M_i, M_j]$ for $i < j$ is induced by ϕ. Then H acts by automorphisms on P, fixing $\mathrm{Z}(P)$ elementwise, and we set $S_n = P \rtimes H$. It is easy to see that $S_n = [S_n, S_n] = S_n^2$. Since $(zx)^2 = x^2$ for every $z \in \mathrm{Z}(P)$ and $x \in S_n$, the number of squares in S_n is no more than $|S_n/\mathrm{Z}(P)| = 4^k \cdot 60$; on the other hand $|S_n| = 2^{k(k-1)/2} \cdot 4^k \cdot 60$. This implies that $f(S_n) \geq (k+3)/6 > n$ if we choose $k > 6n$.

Let me conclude with a little exercise for the reader: if G is any profinite group, then every group in $\mathcal{F}(G_0)$ is isomorphic to a *section* of some group in $\mathcal{F}(G)$ (Hint: apply Theorem 2.6.6 to a suitable finitely generated subgroup of G).

2.7. Probability

Every compact topological group has an invariant measure, the *Haar measure*, unique up to a multiplicative constant. Though quite tricky to construct in general, it is very easy to evaluate in the special case of a profinite group G. Let us write $\mu(X)$ for the measure of a subset X of G, and normalize μ so that $\mu(G) = 1$. If H is an open subgroup of G then each coset Hx of H has the same measure, so

$$\mu(Hx) = |G : H|^{-1}\mu(G) = |G : H|^{-1}.$$

Similarly, $\mu(xH) = |G : H|^{-1}$. As the cosets of open subgroups form a base for the open sets in G, this determines the measure of every open set, and hence also of every closed set. Assuming that G is countably based (i.e. has only countably many open normal subgroups) it is easy to deduce that for any closed subset X of G we have

$$\mu(X) = \inf \frac{|\pi(X)|}{|\pi(G)|} \tag{2.7.1}$$

where π ranges over all the quotient maps $G \to G/N$, N an open normal subgroup.

Now a measure space of measure 1 is a *probability space*: we interpret $\mu(X)$ as the probability that a random element of G belongs to the subset X (note that when $\pi(G)$ is finite, $|\pi(X)|/|\pi(G)|$ is just the proportion of elements of $\pi(G)$ that lie in $\pi(X)$). So we can ask questions about the probability of interesting group-theoretic events; for example, what is the probability that a random k-tuple of elements generates G (topologically)? To make this precise we need to consider the measure on $G^{(k)} = G \times \cdots \times G$, still denoted μ, and define

$$P(G, k) = \mu(X_k)$$

where

$$X_k = \{(x_1, \ldots, x_k) \in G^{(k)} | \langle x_1, \ldots, x_k \rangle = G\}.$$

(Here $\langle S \rangle$ denotes the closed subgroup of G generated by the subset S.) The formula (2.7.1) becomes

$$P(G, k) = \inf P(G/N, k) \tag{2.7.2}$$

where N ranges over all open normal subgroups of G. Obviously, $P(G, k) = 0$ unless G can be generated by k elements. But the converse is not always true. Consider for example the procyclic group $G = \widehat{\mathbb{Z}}$, the profinite completion of the infinite cyclic group $\mathbb{Z}$. Certainly G can be generated by one element. On the other hand, it is easy to see that if $n = p_1^{f_1} \ldots p_r^{f_r}$ then

$$P(\mathbb{Z}/n\mathbb{Z}, k) = \prod_{i=1}^{r} P(\mathbb{Z}/p_i^{f_i}\mathbb{Z}, k) = \prod_{i=1}^{r}\left(1 - \frac{1}{p_i^k}\right)$$

(since a subset Y generates $\mathbb{Z}/p^f\mathbb{Z}$ unless $Y \subseteq p\mathbb{Z}/p^f\mathbb{Z}$). Thus (2.7.2) gives

$$\begin{aligned} P(\widehat{\mathbb{Z}}, k) &= \prod_p \left(1 - \frac{1}{p^k}\right) \\ &= \zeta(k)^{-1} \\ &= \begin{cases} 0 & (k = 1) \\ \frac{6}{\pi^2} & (k = 2) \end{cases}. \end{aligned}$$

The procyclic group $\widehat{\mathbb{Z}}$ is 'only just' a one-generator group: almost all elements do not generate it. On the other hand, a positive proportion – about $3/5$ – of pairs do generate $\widehat{\mathbb{Z}}$.

Avinoam Mann calls a profinite group *positively finitely generated*, or PFG, if $P(G, k) > 0$ for some natural number k. To get some feeling for this property, note that $(x_1, \ldots, x_k)$ belongs to the set X_k defined above if and only if no maximal (open, proper) subgroup of G contains all of $x_1, \ldots, x_k$. That is,

$$G^{(k)} \backslash X_k = \bigcup_{M \in \mathcal{M}} M^{(k)}$$

where $\mathcal{M}$ denotes the set of all maximal subgroups of G. It follows that

$$\begin{aligned} 1 - P(G, k) &= \mu\left(\bigcup_{M \in \mathcal{M}} M^{(k)}\right) \\ &\leq \sum_{M \in \mathcal{M}} \mu(M^{(k)}) \\ &= \sum_{M \in \mathcal{M}} |G : M|^{-k} = \sum_{n \geq 2} m_n(G) n^{-k} \end{aligned}$$

where $m_n(G)$ is the *number of maximal subgroups of index n in G*. Thus $P(G, k)$ is positive if the final sum is less than 1. Suppose for example that the numbers $m_n(G)$ grow at most like a power of n – in this case G is said to have *polynomial maximal subgroup growth*, or PMSG. Then for a certain α we have

$$1 - P(G, k) \leq \sum_{n \geq 2} n^{\alpha - k} = \zeta(k - \alpha) - 1$$

which is less than 1 if $k - \alpha \geq 2$.

It follows that *every profinite group with* PMSG *is* PFG. Since PMSG is a weaker condition than polynomial subgroup growth, we have the corollary that *every profinite group with* PSG *is finitely generated*. This fact can also be seen from the characterization of profinite PSG groups, discussed in §2.4, above; but it is remarkable that it emerges from such a simple probabilistic argument. This simple argument is not reversible, of course; a much more difficult argument, using detailed information about the maximal subgroups of finite simple groups, enabled Mann and Shalev to prove

Theorem 2.7.1 ([MSh]). *A profinite group is positively finitely generated if and only if it has polynomial maximal subgroup growth.*

The class of profinite groups with PMSG is very wide. For example, Borovik, Pyber and Shalev [BPS] have shown that if the profinite group G is finitely generated, then G has PMSG unless G involves *every finite group* as an upper section; also iterated wreath products of finite simple groups, of the type discussed in §2.6 above, have PMSG. So one may say that finitely generated profinite groups have a tendency to be PFG. But the two conditions are certainly not equivalent, since for example a non-abelian finitely generated free profinite group (the profinite completion of a free group) is never PFG.

Probabilistic arguments of the type given above yield all sorts of information. The arguments always take place in the context of a profinite group, but the conclusions sometimes apply to groups in general. I will mention three results, all due to Mann; for the (remarkably simple) proofs, and more discussion of the topic in general, see Chapter 11 of [SG].

1. Let $a_{n,d}(G)$ denote the number of d-generator subgroups of index n in a group G.

Theorem 2.7.2. *Let $m, d \in \mathbb{N}$. Suppose that G is a group that does not involve* $\mathrm{Alt}(m+1)$ *as an upper section. Then there exist C and k, depending only on d and m, such that*

$$a_{n,d}(G) \leq Cn^k$$

for all n.

2. Let $\mathrm{d}(H)$ denote the minimal size of a (topological) generating set for the profinite group H.

Theorem 2.7.3. *Let G be a profinite group with PSG. Then there exists a constant C such that*

$$\mathrm{d}(H) \leq C\sqrt{\log |G:H|}$$

for every open subgroup H of G.

3. Let $h(n, r)$ denote the number of (isomorphism types of) groups of order n having a finite presentation with r relations.

Theorem 2.7.4. *Let p be a prime and $r \in \mathbb{N}$. Then*

$$h(p^k, r) = o(p^{kr}) \qquad \text{as } k \to \infty.$$

Many other results and problems are given in [M1] and [M2]. One of the most intriguing of those was the following question: *does every open subgroup of a PFG group have PFG*? In very recent and significant work [J-ZP], Jaikin-Zapirain and Pyber have proved that the answer is *yes*. They do this by providing a detailed characterization of groups with PMSG in terms of the structure of their finite quotients. As a consequence of their methods they also establish the following extension of Theorem 2.7.4: *there is an absolute constant c such that*

$$h(n, r) \leq n^{cr} \text{ for all } n.$$

2.8. Other topics

2.8.1. *The congruence subgroup problem*

I have referred to the 'congruence subgroup property' in several of the preceding sections. Recall that a subgroup Γ in some profinite group G is said to have the CSP if the topology of G induces on Γ its own profinite topology. This is equivalent to saying that the natural map $\widehat{\Gamma} \to G$ is injective, or in down-to-earth terms that *every subgroup of finite index in* Γ *contains* $\Gamma \cap N$ *for some open subgroup* N *of* G. This terminology originates in a very classical problem: what are the subgroups of finite index in $\Gamma = \mathrm{SL}_n(\mathbb{Z})$? There are some obvious ones: for an integer $m \neq 0$ the *principal congruence subgroup* mod m is

$$\begin{aligned}\Gamma(m) &= \{g \in \Gamma \mid g_{ij} \equiv \delta_{ij} \ (\mathrm{mod}\ m) \text{ for } 1 \leq i, j \leq n\} \\ &= \ker(\Gamma \to \mathrm{SL}_n(\mathbb{Z}/m\mathbb{Z})),\end{aligned}$$

and one calls any subgroup of Γ that contains $\Gamma(m)$ for some $m \neq 0$ a *congruence subgroup*. Evidently, the congruence subgroups have finite index in Γ, and the problem is: are there any others? This was solved in the 1960s by Mennicke and Bass, Lazard and Serre: they proved that the answer is 'no' when $n \geq 3$; as for the case $n = 2$, it had been known since the 19th century that $\mathrm{SL}_2(\mathbb{Z})$ has an abundance of non-congruence subgroups of finite index.

If every subgroup of finite index is a congruence subgroup, the group Γ is said to have the *congruence subgroup property*. We see that this is a special case of the previous definition if we consider Γ as a subgroup of the profinite group

$$\widetilde{\Gamma} = \mathrm{SL}_n(\widehat{\mathbb{Z}}),$$

so the congruence subgroup problem can be formulated as: is the natural map $\widehat{\Gamma} \to \widetilde{\Gamma}$ injective?

Now $\mathrm{SL}_n(\mathbb{Z})$ is just the most familiar example of the important class of *S-arithmetic groups*, and the analogous question applies to all such groups. I will not define these here in full generality: for a comprehensive account see the book [PR]. Typical examples are groups of the form $\Gamma = \mathfrak{G}(\mathbb{Z}_S)$ where $\mathfrak{G}$ is an algebraic matrix group defined over $\mathbb{Q}$, S is a finite set of primes and $\mathbb{Z}_S = \mathbb{Z}\left[\frac{1}{p};\ p \in S\right]$. The congruence subgroup problem now becomes: determine the kernel $C(\mathfrak{G}, S)$ of the natural map

$$\widehat{\Gamma} \to \mathfrak{G}(\widehat{\mathbb{Z}_S}).$$

This group $C(\mathfrak{G}, S)$ is called the *congruence kernel*. It was observed by Serre that the natural dichotomy seems to be between those groups whose congruence kernel is *finite* and those for which it is *infinite*, and following his insight it is usual now to say that Γ has the CSP if $C(\mathfrak{G}, S)$ is *finite* (note that according to the original definition, we would require $C(\mathfrak{G}, S) = 1$). The following very general conjecture was made by Serre:

Conjecture. Let $\mathfrak{G}$ be a simple simply connected algebraic group over a global field k and let S be a finite set of places of k. Then (under certain natural assumptions) the S-arithmetic group $\mathfrak{G}(\mathcal{O}_S)$ has the CSP if and only the S-rank of $\mathfrak{G}$ is at least 2.

Here, $\mathcal{O}_S$ denotes the ring of 'S-integers' of k; the 'S-rank' of $\mathrm{SL}_n(\mathbb{Z}_S)$, for example, is equal to $n - 1 + |S|$. This conjecture has been proved in the majority of cases, but some hard problems remain open: see for example [Ra].

An interesting recent development relates the congruence subgroup property of Γ to purely group-theoretic properties of Γ, such as its subgroup growth and its index growth. These results are due in the main to Platonov, Rapinchuk and Lubotzky; for a detailed account of some of them see Chapter 7 of [SG].

2.8.2. *Profinite presentations*

By a *presentation* of a group G is meant an epimorphism $\pi : F \to G$, where F is a free group, together with a specific choice of a set X of free generators for F and a set R of generators for the kernel $\ker \pi$ as a normal subgroup of F. It is usual to write

$$G = \langle X; R \rangle,$$

where R is a set of words on the alphabet X, and to interpret the symbols in X as generators of G that satisfy the relations $w(X) = 1$ for all $w \in R$. For profinite groups, it is natural to consider instead epimorphisms from a *free profinite* group. When X is a finite set (the only case we consider here), the free profinite group $\widehat{F}(X)$ on X is just the profinite completion of the free group on X, and it has the expected universal property with respect to continuous mappings from X into profinite groups. A *profinite presentation* of G is thus a *continuous* epimorphism $\pi : \widehat{F}(X) \to G$, together with a choice R of generators for $\ker \pi$ as a *closed* normal subgroup of $\widehat{F}(X)$. The elements of R need no longer be words in the generators X: in general they are 'profinite words', that is, limits of convergent sequences of ordinary words. But we still write

$$G = \langle X; R \rangle$$

to indicate such a profinite presentation (as long as the context makes it clear which kind of presentation is meant).

The usefulness of this concept lies in the simple observation that if $\Gamma = \langle X; R \rangle$ is an ordinary presentation of some abstract group Γ, then $G = \langle X; R \rangle$ is a profinite presentation of the profinite completion $G = \widehat{\Gamma}$. Given information about a presentation of Γ, we can therefore interpret it as information about $\widehat{\Gamma}$; profinite group theory may then yield conclusions about $\widehat{\Gamma}$, which in turn gives us information about Γ. This will be illustrated below. First I want to mention a celebrated open problem.

Write $\mathrm{d}(G)$ to denote the minimal number of generators required for a group G (topological generators in the profinite context), and call $G = \langle X; R \rangle$ a 'minimal presentation' (in either case) if $|X| = \mathrm{d}(G)$. The minimal number of relations required for some minimal presentation of G (in either context) is denoted $\mathrm{t}(G)$. Now suppose that Γ happens to be a *finite* group. In this case, of course, $\widehat{\Gamma} = \Gamma$, and we may interpret the expression $\Gamma = \langle X; R \rangle$ either as an ordinary presentation or as a profinite presentation. Since the topology on Γ is discrete, a set X generates Γ if and only if it generates Γ topologically. But the topology on $\widehat{F}(X)$ is by no means discrete: just for now, let us understand $\mathrm{t}(\Gamma)$ in the abstract sense, and write $\mathrm{t}(\widehat{\Gamma})$ for the minimal number of relations in a minimal profinite presentation of Γ.

Problem. Let Γ be a finite group. Is $\mathrm{t}(\widehat{\Gamma})$ necesarily equal to $\mathrm{t}(\Gamma)$?

(If r ordinary relations suffice to define Γ, then the same relations also define Γ as a profinite group; but it is conceivable that Γ could be defined using a *smaller* number of *profinite* relations.) For some discussion, and alternative formulations, of this problem see §2.3 of [SG] (*Remark* on page 48).

Two striking applications of the philosophy outlined above were made by Lubotzky. The first uses *pro-p presentations* rather than profinite ones: these are defined in exactly the same way, using free pro-p groups in place of free profinite groups.

1. The famous theorem of **Golod and Shafarevich** asserts that if G is a finite p-group, then

$$\mathrm{t}(G) \geq \frac{\mathrm{d}(G)^2}{4} \tag{2.8.1}$$

(this is true in either interpretation of the symbols, abstract or pro-p). This was generalized (by Koch and Lubotzky, using Lazard's theory) to the case of any *p-adic analytic pro-p group G* (with $\mathrm{d}(G)$ and $\mathrm{t}(G)$ now defined in terms of pro-p presentations, of course). This has consequences for any abstract group Γ whose pro-p completion is such a group G; in general, $\mathrm{d}(\Gamma)$ may be strictly larger than $\mathrm{d}(G)$, but if, for example, Γ is nilpotent then there exist primes p such that $\mathrm{d}(\Gamma) = \mathrm{d}(\widehat{\Gamma}_p)$, and one may deduce

Theorem 2.8.1. *Let Γ be a finitely generated non-cyclic nilpotent group. Then* $\mathrm{t}(\Gamma) \geq \mathrm{d}(\Gamma)^2/4$.

This is a direct generalization of the original Golod–Shafarevich theorem to infinite groups. For details of the argument see [DDMS], Interlude D. By further generalizing the Golod–Shafarevich theorem to a larger class of pro-p groups, J. S. Wilson established a result of still wider applicability (it includes all finitely generated soluble groups, for example):

Theorem 2.8.2 ([W1]). *Let Γ be a group which has no infinite p-torsion residually finite quotient, for any prime p. Suppose that Γ has a presentation with n generators and r relations. Then*

$$r \geq n + \frac{d^2 - 1}{4} - d$$

where $d = \mathrm{d}(\Gamma^{\mathrm{ab}})$.

Here $\Gamma^{\mathrm{ab}} = \Gamma/\Gamma'$ denotes the abelianization of Γ; this appears because $\mathrm{d}(\Gamma^{\mathrm{ab}})$ (unlike $\mathrm{d}(\Gamma)$) can be recognised as $\mathrm{d}(\widehat{\Gamma}_p)$ for a suitable prime p.

Lubotzky was concerned with groups that are very far from soluble. Let Γ be an *arithmetic lattice* in $\mathrm{SL}_2(\mathbb{C})$ – examples include groups like $\mathrm{SL}_2(\mathcal{O})$ where $\mathcal{O}$ is the ring of integers in an imaginary quadratic field, but there are more mysterious ones. It is fairly easy to see that if Γ has the congruence subgroup property then its pro-p completion $\widehat{\Gamma}_p = G$ is p-adic analytic, and hence satisfies (2.8.1); moreover, the same holds for the pro-p completion of every subgroup Δ of finite index in Γ. From this it may be deduced that

$$|X| - |R| \leq \mathrm{d}_p(\Delta) - \frac{\mathrm{d}_p(\Delta)^2}{4}$$

for every finite presentation $\Delta = \langle X;\, R\rangle$, where

$$\mathrm{d}_p(\Delta) = \mathrm{d}(\widehat{\Delta}_p) = \mathrm{d}(\Delta/\Delta^p[\Delta,\Delta]).$$

On the other hand, according to a theorem of Epstein each such Δ has a presentation $\langle X;\, R\rangle$ for which $|R| \le |X|$ (assuming, as we may, that Δ is torsion-free). Hence $\mathrm{d}_p(\Delta) \le 4$. Now the theory of linear groups shows that if the numbers $\mathrm{d}_2(\Delta)$ are *bounded* as Δ ranges over all the subgroups of finite index in Γ, then Γ must have a soluble subgroup of finite index. This is certainly not the case here, so we have

Theorem 2.8.3 ([Lu1]). *No arithmetic lattice in* $\mathrm{SL}_2(\mathbb{C})$ *satisfies the congruence subgroup property.*

This establishes many of the 'negative' cases of Serre's conjecture, stated in the preceding subsection. The method has been generalized by Lubotzky to obtain

Theorem 2.8.4. *Let* Γ *be any lattice in* $\mathrm{SL}_2(\mathbb{C})$. *Then* Γ *has subgroup growth of type at least* $n^{(\log n)^{2-\varepsilon}}$ *for every* $\varepsilon > 0$.

A *lattice* is a discrete subgroup of finite co-volume. Since the *congruence subgroup growth* of any arithmetic group is at most of type $n^{\log n/\log\log n}$, this shows that the congruence subgroup property fails here in a dramatic way: the subgroups of finite index vastly outnumber the congruence subgroups as the index goes to infinity. For details of the proof, and many other cases, see Chapter 7 of [SG].

Returning to profinite groups, or rather pro-p groups, the most powerful generalization of the Golod–Shafarevich theorem was obtained by Zelmanov:

Theorem 2.8.5 ([Z]). *Let* G *be a non-procyclic finitely generated pro-*p *group with a minimal pro-*p *presentation* $G = \langle X;\, R\rangle$. *Then either* $|R| \ge |X|^2/4$ *or else* G *contains a closed subgroup that is a non-abelian free pro-*p *group.*

2. Let $f(n, d)$ denote the number of (isomorphism types of) d-generator groups of order n. Establishing a conjecture of Mann, Lubotzky proved

Theorem 2.8.6. *For every* n *and* d *we have*

$$f(n, d) \le n^{2(d+1)\lambda(n)}.$$

Here $\lambda(n) = \sum l_i$ where $n = \prod p_i^{l_i}$ is the factorization of n into prime-powers. This is deduced from the following theorem: *every finite simple group of order* n *has a profinite presentation with* 2 *generators and at most* $2\lambda(n)$ *relations*. It is conjectured that this remains true if the word 'profinite' is omitted, and this has been proved in most cases. But it is in general easier to get at a profinite presentation than at an abstract presentation: roughly speaking, if $N = \ker\pi$ in our original notation, then the number of *profinite* relations needed for a presentation $\pi : F \to G$ can be detected in the 'relation module' $N/[N, N]$, whereas the number of 'ordinary' relations depends on the structure of N itself as an F-operator group. For details, see §2.3 of [SG].

2.8.3. *Profinite trees*

A large part of combinatorial group theory deals with the properties of generalized free products and HNN extensions. A powerful unified framework for studying such constructions is the Bass-Serre theory of groups acting on trees. In recent years, an analogous theory has been developed of profinite groups acting on 'profinite trees', largely due to the work of Melnikov, Ribes and P. A. Zalesskii. As well as providing a basis for the theory of generalized free products in the profinite category, this has found a number of applications to abstract free groups and free products; a typical example is Theorem 2.2.8 mentioned in §2.2, above.

This is a significant chapter in 'pure' profinite group theory, with solid achievements but also presenting a number of challenging open problems. However, it is beyond my competence to present anything like an adequate account of it. Detailed expositions of the theory are given in [RZ1] (for pro-p groups) and the forthcoming book [RZ3]; for various specific applications, see the papers [RZ4], [RZ5], [RZ6] and [RSZ].

References

[BG] L. Bartholdi, R. I. Grigorchuk and Z. Sunić, Branch groups, in Handbook of Algebra III, ed. M. Hazewinkel, North-Holland, Amsterdam (2003).

[B] N. Boston, p-adic Galois representations and pro-p Galois groups, Chapter 11 in [NH].

[BCRS] G. Baumslag, F. Cannonito, D. J. S. Robinson and D. Segal, The algorithmic theory of polycyclic-by-finite groups, *J. Algebra*, **142** (1991) 118–149.

[BMP] A. Balog, A. Mann and L. Pyber, Polynomial index growth groups, *Int. J. Algebra and Computation*, **10** (2000) 773–782.

[BPS] A. Borovik, L. Pyber and A. Shalev, Maximal subgroups in finite and profinite groups, *Trans. Amer. Math. Soc.*, **348** (1996) 3745–3761.

[CG] J-P. Serre, Galois Cohomology, Springer Verlag, Berlin-Heidelberg (1997).

[DDMS] J. D. Dixon, M. P. F. du Sautoy, A. Mann and D. Segal, Analytic pro-p Groups, 2nd edition, Cambridge Studies in Advanced Maths., **61**, Cambridge Univ. Press, Cambridge (1999).

[dS1] M. P. F. du Sautoy, Finitely generated groups, p-adic analytic groups and Poincaré series, *Annals of Math.*, **137** (1993) 639–670.

[dS2] M. P. F. du Sautoy, Counting p-groups and nilpotent groups, *Publ. Math. IHES*, **92** (2000) 63–112.

[dSF] M. P. F. du Sautoy and I. Fesenko, Where the wild things are: ramification groups and the Nottingham group, Chapter 10 in [NH].

[dSS] M. P. F. du Sautoy and D. Segal, Zeta functions of groups, Chapter 9 in [NH].

[dSW] M. P. F. du Sautoy and L. Woodward, Zeta Functions of Groups and Rings, Lecture Notes in Math., **1925**, Springer-Verlag, Berlin Heidelberg (2008).

[E1] B. Eick, Orbit-stabilizer problems and computing normalizers for polycyclic groups, *J. Symbolic Comp.*, **34** (2002) 1–19.

[E2] B. Eick, Computing with infinite polycyclic groups, Groups and computation, III (Columbus, Ohio 1999), pp. 139–154, Ohio State Univ. Math. Res. Inst. Publ. 8, de Gruyter, Berlin (2001).

[EL] B. Eick and C.R. Leedham-Green, On the classification of prime-power groups by coclass, *Bull. London Math. Soc.*, to apear.

[FJ] M. D. Fried and M. Jarden, Field arithmetic, Ergebnisse der Math. (3), **11**, Springer-Verlag, Berlin–Heidelberg (1986).

[G] R. I. Grigorchuk, Just infinite branch groups, Chapter 4 in [NH].

[GPS] F. J. Grunewald, P. F. Pickel and D. Segal, Polycyclic groups with isomorphic finite quotients, *Annals of Math.*, **111** (1980) 155–195.

[GSS] F. J. Grunewald, D. Segal and G. C. Smith, Subgroups of finite index in nilpotent groups, *Invent. Math.*, **93** (1988) 185–223.

[H] D. F. Holt, Enumerating perfect groups, *J. London Math. Soc.*, **39** (1989) 67–78.

[J-Z1] A. Jaikin-Zapirain, Zeta function of representations of compact p-adic analytic groups, *J. Amer. Math. Soc.*, **19** (2006) 91–118.

[J-Z2] A. Jaikin-Zapirain, On the verbal width of finitely generated pro-p groups, *Revista Mat. Iberoamericana*, **24** (2008) 617–630.

[J-ZP] A. Jaikin-Zapirain and L. Pyber, Random generation of finite and profinite groups and group enumeration, to appear.

[KN] M. Kassabov and N. Nikolov, Cartesian products as profinite completions, *International Math. Research Notices*, **2006** (2006) Article ID 72947.

[KN2] Kassabov and Nikolov, Generation of polycyclic groups, *J. Group Theory*, **12** (2009) 567–577.

[L] M. Lazard, Groupes analytiques p-adiques, *Publ. Math. IHES*, **26** (1965) 389–603.

[LL] M. Larsen and A. Lubotzky, Representation Growth for Linear Groups, arXiv:math.GR/0607369.

[LGM1] C. R. Leedham-Green and S. McKay, The structure of groups of prime-power order, LMS Monographs **27**, Oxford Univ. Press, Oxford (2002).

[LGM2] C. R. Leedham-Green and S. McKay, On the classification of p-groups and pro-p groups, Chapter 2 in [NH].

[Lu1] A. Lubotzky, Group presentation, p-adic analytic groups and lattices in $SL_2(\mathbb{C})$, *Ann. of Math. (2)*, **118** (1983) 115–130.

[Lu2] A. Lubotzky, A group-theoretic characterization of linear groups, *J. Algebra*, **113** (1988) 207–214.

[LM] A. Lubotzky and A. Mann, Powerful p-groups, I: Finite groups; II: p-adic analytic groups, *J. Algebra*, **105** (1987) 484–505, 506–515.

[LMS] A. Lubotzky, A. Mann and D. Segal, Finitely generated groups of polynomial subgroup growth, *Israel J. Math.*, **82** (1993) 363-371.

[LS] R. C. Lyndon and P. E. Schupp, Combinatorial group theory, Springer-Verlag, Berlin-Heidelberg-New York (1977).

[LW] P. A. Linnell and D. Warhurst, Bounding the number of generators of a polycyclic group, *Archiv der Math.*, **37** (1981) 7–17.

[M1] A. Mann, Positively finitely generated groups, *Forum Math.*, **8** (1996) 429–459.

[M2] A. Mann, Some applications of probability in group theory, Groups: topological, combinatorial and arithmetic aspects, ed. T. W. Mü ller, LMS Lect. Note Series **311**, CUP, Cambridge (2004).

[MS1] A. Mann and D. Segal, Uniform finiteness conditions in residually finite groups, *Proc. London Math. Soc. (3)*, **61** (1990) 529–545.

[MS2] A. Mann and D. Segal, Subgroup growth: some current developments, in Infinite Groups 94, eds. de Giovanni and Newell, W. de Gruyter (1995).

[MSh] A. Mann and A. Shalev, Simple groups, maximal subgroups and probabilistic aspects of profinite groups, *Israel J. Math.*, **96** (1996), 449–468.

[NH] M. P. F. du Sautoy, D. Segal and A. Shalev (ed.), New horizons in pro-p groups, *Progress in Math.*, **184** Birkhäuser Boston (2000).

[NS1] N. Nikolov and D. Segal, Finite index subgroups in profinite groups, *C. R. Acad. Sci. Paris, Ser. I*, **337** (2003) 303–308.

[NS2] N. Nikolov and D. Segal, On finitely generated profinite groups, I: strong completeness and uniform bounds. II: products in quasisimple groups, *Annals of Math.*, **165** (2007) 171–238, 239–273.

[NS3] N. Nikolov and D. Segal, Direct products and profinite completions, *J. Group Theory*, **10** (2007) 789–793.

[PR] V. P. Platonov and A. S. Rapinchuk, Algebraic groups and number theory, Academic Press, San Diego (1994).

[P] L. Pyber, Groups of intermediate subgroup growth and a problem of Grothendieck, *Duke Math. J.*, **121** (2004) 169–188.

[PS] L. Pyber and D. Segal, Finitely generated groups with polynomial index growth, *J. reine angewandte Math.*, **612** (2007) 173–211.

[Ra] A. S. Rapinchuk, The congruence subgroup problem, Algebra, K-theory, groups, and education (New York, 1997), 175–188, *Contemp. Math.*, **243**, Amer. Math. Soc., Providence, RI (1999).

[R] V. A. Romankov, Width of verbal subgroups in solvable groups, *Algebra i Logika*, **21** (1982) 60–72 (Russian); *Algebra and Logic*, **21** (1982) 41–49 (English).

[RZ1] L. Ribes and P. A. Zalesskii, Pro-p trees and applications, Chapter 3 in [NH].

[RZ2] L. Ribes and P. A. Zalesskii, Profinite groups, Ergebnisse der Math. **40**, Springer, Berlin – Heidelberg (2000).

[RZ3] Ribes and Zalesskii, Profinite trees, Springer, Berlin (to appear).

[RZ4] Ribes and Zalesskii, On the profinite topology on a free group, *Bull. London Math. Soc.*, **25** (1993) 37–43.

[RZ5] Ribes and Zalesskii, The pro-p topology of a free group and algorithmic problems in semigroups, *Int. J. Algebra and Comput.*, **4** (1994) 359–374.

[RZ6] Ribes and Zalesskii, Profinite topologies in free products of groups, *Int. J. Algebra and Comput.*, **14** (2004) 751–772.

[RSZ] L. Ribes, D. Segal and P. A. Zalesskii, Conjugacy separability and free products with cyclic amalgamation, *J. London Math. Soc. (2)*, **57** (1998) 609–628.

[S] D. Segal, Polycyclic groups, Cambridge Univ. Press, Cambridge (1983). (Reprinted in paperback, 2005.)

[S1] D. Segal, Decidable properties of polycyclic groups, *Proc. London Math. Soc. (3)*, **61** (1990) 497–528.

[S2] D. Segal, Variations on polynomial subgroup growth, *Israel J. Math.*, **94** (1996) 7–19.

[S3] D. Segal, Subgroups of finite index in soluble groups II, in *Groups St Andrews 1985*, LMS Lect. note series **121**, pp. 315–319, CUP, Cambridge (1986).

[S4] D. Segal, The finite images of finitely generated groups, *Proc. London Math. Soc. (3)* **82** (2001) 597–613.

[Sh] A. Shalev, The structure of finite p-groups: effective proof of the coclass conjectures, *Invent. Math.*, **115** (1994) 315–345.

[SG] A. Lubotzky and D. Segal, Subgroup Growth, Progress in Math., **212**, Birkhäuser, Basel (2003).

[SSh] D. Segal and A. Shalev, Profinite groups with polynomial subgroup growth, *J. London Math. Soc. (2)*, **55** (1997) 320–334.

[V] C. Voll, Functional equations for zeta functions of groups and rings, arXiv:math.GR/0612511; Annals of Math., to appear.

[W] D. Segal, Words, London Math. Soc. Lecture Notes Series, **361**, Cambridge Univ. Press, Cambridge, 2009.

[W1] J. S. Wilson, Finite presentations of pro-p groups and discrete groups, *Invent. Math.*, **105** (1991) 177–183.

[W2] J. S. Wilson, Profinite groups, London Math. Soc. Monographs (n.s.) **19**, Clarendon Press, Oxford (1998).

[Wi] D. T. Wise, A residually finite version of Rips's construction, *Bull. London Math. Soc.*, **35** (2003) 23–29.

[Z] E. Zelmanov, On groups satisfying the Golod-Shafarevich condition, Chapter 7 in [NH].

Proc. Inst. Wrk.shps. Int. Conf. – Essays in Geometric Group Theory No. 9, 2009, pp. 61–77.

Ultrapowers of Lie Groups, a Question of Gromov, and the Margulis Conjecture

Linus Kramer and Katrin Tent

Math. Institut, Univ. Münster, Einsteinstr. 62, 48149 Münster, Germany
e-mail: linus.kramer@math.uni-muenster.de; tent@math.uni-muenster.de

Using ultrapowers of Lie groups and nonstandard techniques, we obtain a short proof of the Margulis conjecture: in the mid-seventies, Margulis asked whether for any quasi-isometry of Riemannian symmetric spaces there is an isometry at bounded distance from it (all notions are explained below). This was a reasonable conjecture, as Mostow had just proved his famous Rigidity Theorem [20], saying that any such quasi-isometry coming from the action of a uniform lattice is accompanied by an isometry. Mostow's proof relied very much on Tits' results about spherical buildings and what one would call now the large-scale geometry of Riemannian symmetric spaces. In a sense, Margulis asked for a lattice-free proof of Mostow Rigidity. However, the lattice is absolutely crucial in Mostow's argument, and the question remained open until it was recently settled by Kleiner–Leeb [14], using tools from differential geometry, Euclidean buildings (in their sense) and asymptotic cones (in the sense of Gromov) of symmetric spaces.

A special, but may be instructive way to view this problem is as follows. Suppose that $Q \subseteq X$ is a countable discrete subset of the Riemannian symmetric space X, such that for some fixed real number r, the r-balls about the elements of Q cover X: can one tell the symmetric space X just from the discrete metric space Q (or any discrete metric space Q' quasi-isometric to Q)?

Our approach relies on the fact that every Riemannian symmetric space X can be presented algebraically as a quotient $X = G/K$ of some semisimple real Lie group G, the isometry group of X, by one of its maximal compact subgroups K. We use the model-theoretic techniques of *asymptotic cones* introduced by van den Dries and Wilkie [32] and Gromov, and define new structures which we call *ultracones*, both of which are ultrapower constructions reflecting the large-scale structure of metric spaces, yielding powerful functors on the category of metric spaces. The idea is that by considering an ultrapower $^*X$ of X, we obtain essentially the same space over a nonstandard extension $^*\mathbb{R}$ of the real numbers $\mathbb{R}$. For the asymptotic cone, one re-scales the metric and identifies (afterwards) points at infinitesimal distance, while for the ultracone, one identifies points at finite distance. In both cases, the quasi-isometry becomes a bi-Lipschitz homeomorphism of the resulting spaces, which turns out to preserve a certain combinatorial structure, showing the isometry groups to be isomorphic. This implies that the spaces are isometric, and using nonstandard techniques, we then show that the quasi-isometry has bounded distance from a unique isometry.

The construction of van den Dries and Wilkie generalizes the notion of Gromov–Hausdorff convergence of sequences of metric spaces, enforcing convergence. Later, Gromov translated it back again into a non-model-theoretic phrasing [12]. Our point of view is much closer to the original paper [32]; in this way, it is easier to keep track of the various identifications which occur during the collapsing. In particular, we obtain an explicit group-theoretic model for the nondiscrete affine building arising from X as an asymptotic cone (resp. ultracone). Assuming ($\neg$CH) this model then allows us to answer a question of Gromov about asymptotic cones of finitely presented groups extending results in [28].

This report is meant as a survey without detailed proofs, for which we refer the reader to the papers [17,18].

Introduction

A (not necessarily continuous) map $f : X \longrightarrow X'$ between metric spaces is called an (L, C)-*quasi-isometry* if there exist constants $L \geq 1$ and $C \geq 0$ such that

$$L^{-1} \cdot d(x, y) - C \leq d'(f(x), f(y)) \leq L \cdot d(x, y) + C$$

holds for all $x, y \in X$, and such that for every $x' \in X'$ there exists some $x \in X$ with $d'(f(x), x') \leq C$. Special instances of quasi-isometries are

(1) *isometries* (where $(L, C) = (1, 0)$)
(2) *bi-Lipschitz maps* (where $C = 0$)
(3) *coarse isometries* (where $L = 1$).

Let X be a Riemannian symmetric space of noncompact type, and let $G = I(X)^\circ$ denote its connected isometry group. Then G is a (finite extension of) a semi-simple real Lie group acting transitively on X (see Section 5). The point stabilizer of this action is a maximal compact subgroup K of G, so that we may identify $X = G/K$. The relevant geometric structure of X like the metric and the concept of maximal flats can be expressed in terms of this algebraic presentation (see Section 3.5). This point of view allows us to use model theoretic machinery.

The Margulis conjecture may be stated as follows:

Margulis Conjecture. Let X and X' be Riemannian symmetric spaces of noncompact type (without rank 1 factors). If $f : X \longrightarrow X'$ is a quasi-isometry, then X and X' are isometric, and there exists an isometry $\bar{f} : X \longrightarrow X'$ at bounded distance from f.

Kleiner–Leeb [14] proved the Margulis conjecture in 1997; they used Gromov's technique of asymptotic cones [12] rather than the framework developed by van den Dries and Wilkie. Kleiner–Leeb showed that the asymptotic cone of a Riemannian symmetric space is an affine $\mathbb{R}$-building, and they used this building and its geometry systematically in their work [14]. (Note that they work with a class of metric spaces they call *Euclidean buildings*; their system of axioms is different from Tits' *systèmes d'appartements* [31]. Parreau [22] proved that their spaces are a subcase of Tits' systèmes d'appartements.)

However, it does not follow from their paper which building the asymptotic cone of a given Riemannian symmetric space really is. We give a general group-theoretic construction of affine $\mathbb{R}$-buildings, and more generally, of affine Λ-buildings, associated to semisimple Lie groups over nonarchimedean real closed fields. Bennett's affine Λ-buildings (which include both the Kleiner Leeb Euclidean buildings and Tits' *systèmes d'appartements* as a strict subcase) are crucial for our argument. They simplify in particular the construction of the accompanying isometry in a drastic way. Using model theory we are able to identify explicitly the building arising as the asymptotic cone. This also allows us to simplify the proof of the Margulis conjecture significantly. Furthermore, with this information we can answer a question of Gromov (see Section 3.7); our proof also sheds some light on the algebraic background of the Margulis conjecture.

3.1. Λ-metric spaces

Throughout, $(\Lambda, +, \leq)$ is an ordered (and hence torsion free) abelian group; everything we need about these groups can be found in [24]. As usual, we define $|\lambda| = \max\{\pm\lambda\}$.

3.1.1. We use the following slight generalization of a metric space.

A Λ-*pseudometric* on a set X is a function $d : X \times X \longrightarrow \Lambda$ which satisfies the usual axioms of a pseudometric for all $x, y, z \in X$:

$$d(x,x) = 0 \quad d(x,y) = d(y,x) \geq 0 \quad d(x,y) + d(y,z) \geq d(x,z).$$

If $d(x, y) = 0$ implies that $x = y$, then the Λ-pseudometric is called a Λ-*metric*. A Λ-pseudometric defines a topology (and a uniformity) on X in the usual way, which is Hausdorff if and only if d is a Λ-metric. For $\Lambda \subseteq \mathbb{R}$ we have of course the traditional version of a real valued (pseudo)metric.

3.1.2. Apart from the Riemannian symmetric spaces (which are metric spaces in the traditional sense) we will be interested in the following class of examples.

Let Γ be a group with finite generating set S and assume that S is closed under inverses. Then for any $g \in \Gamma \backslash \{1\}$, we define $|g|$ to be the *word length* of g, i.e. $|g| = \min\{n \mid g = x_1 \ldots x_n,\ x_i \in S\}$ (and $|1| = 0$). Then one can check that $d(g,h) = |g^{-1}h|$ for $g, h \in \Gamma$ defines a Γ-invariant $\mathbb{Z}$-valued metric, the *word metric* on Γ. If S' is another finite generating set, and d' its corresponding metric, then the identity map can be shown to be a quasi-isometry from (Γ, d) to (Γ, d'), see [8] I.8.17(2).

3.1.3. Recall that the ordered abelian group Λ is *archimedean* if the following holds: for any two elements $x, y > 0$, there exists an $n \in \mathbb{N}$ such that $nx \geq y$. An ordered abelian group is archimedean if and only if its only o-convex (*order convex*) subgroups are 0 and Λ. Any archimedean Λ admits an o-embedding into $(\mathbb{R}, +)$, which is unique up to a scaling factor; thus, our concept of a Λ-metric is broader than the usual concept of a metric only in the nonarchimedean case.

3.1.4. In the nonarchimedean case, suppose that $\Omega \subseteq \Lambda$ is an o-convex subgroup, and that $d : X \times X \longrightarrow \Lambda$ is a Λ-metric. Put

$$x \approx_\Omega y \quad \text{if} \quad d(x,y) \in \Omega.$$

This is an equivalence relation on X, and we put $X_x^\Omega = \{y \in X \mid x \approx_\Omega y\}$. The composite pseudometric $X \times X \xrightarrow{d} \Lambda \longrightarrow \Lambda/\Omega$ induces a Λ/Ω-metric on the set

$$\{X_x^\Omega \mid x \in X\} = X/\Omega$$

of $\approx_\Omega$-equivalence classes in X. In other words, we identify points whose distance is in Ω. This generalizes the well-known process of making a pseudometric into a metric (the case $\Omega = 0$).

3.1.5. Given two elements $\alpha, \beta \geq 0$ in Λ, we write $\alpha \gg \beta$ if $\alpha > n\beta$ holds for all $n \in \mathbb{N}$. Let $\alpha \in \Lambda$ be a positive element ($\alpha > 0$), and let

$$\Lambda^{\langle\alpha\rangle} = \{\lambda \in \Lambda \mid |\lambda| \leq n\alpha \text{ for some } n \in \mathbb{N}\} \text{ and } \Lambda_{\langle\alpha\rangle} = \{\lambda \in \Lambda \mid |\lambda| \ll \alpha\}.$$

Then $\Lambda_{\langle\alpha\rangle}$ is a maximal o-convex subgroup of $\Lambda^{\langle\alpha\rangle}$ and the quotient

$$\Lambda^{(\alpha)} = \Lambda^{\langle\alpha\rangle}/\Lambda_{\langle\alpha\rangle}$$

is archimedean; there exists a unique o-homomorphism $\phi : \Lambda^{\langle\alpha\rangle} \longrightarrow \mathbb{R}$ with $\phi(\alpha) = 1$, whose kernel is precisely $\Lambda_{\langle\alpha\rangle}$,

$$0 \longrightarrow \Lambda_{\langle\alpha\rangle} \longrightarrow \Lambda^{\langle\alpha\rangle} \longrightarrow \mathbb{R}.$$

3.1.6. We can make a very similar construction with any Λ-metric space X. Choose a basepoint $o \in X$ and let

$$X_o^{\langle\alpha\rangle} = \{x \in X \mid x \approx_{\Lambda^{\langle\alpha\rangle}} o\} = \{x \in X \mid d(x,o) \leq n\alpha \text{ for some } n \in \mathbb{N}\}.$$

The composite $\phi \circ d : X_o^{\langle\alpha\rangle} \times X_o^{\langle\alpha\rangle} \longrightarrow \mathbb{R}$ is an $\mathbb{R}$-pseudometric, and induces an $\mathbb{R}$-metric on the quotient

$$X_o^{(\alpha)} = X_o^{\langle\alpha\rangle}/\Lambda_{\langle\alpha\rangle}.$$

3.1.7. Let $L : \Lambda \longrightarrow \Lambda$ be a central o-automorphism and let $C \in \Lambda$ be nonnegative. A (not necessarily continuous) map $f : X \longrightarrow X'$ of Λ-metric spaces (X,d) and (X',d') is an (L,C)-*quasi-isometry* if

$$L^{-1}d(x,y) - C \leq d'(f(x), f(y)) \leq Ld(x,y) + C$$

holds for all $x, y \in X$, and if for every $x' \in X'$, there exists an $x \in X$ such that $d'(x', f(x)) \leq C$. Two quasi-isometries f, g are *equivalent* if they have bounded distance, i.e. If there exists a constant $D \in \Lambda$ such that $d'(f(x), g(x)) \leq D$ holds for all $x \in X$. It is not difficult to see that each quasi-isometry f has a *quasi-inverse* h, i.e. there exists a quasi-isometry $h : X' \longrightarrow X$ such that $f \circ h$ and $h \circ f$ have bounded distances from the identity. The collection of all self-quasi-isometries of X, factored by equivalence, is a group $\mathcal{QI}(X)$, the *quasi-isometry group of* X. There is a natural map from the isometry group $I(X)$ to the quasi-isometry group $\mathcal{QI}(X)$ which is in general neither surjective nor injective, see [8] I.8.16.

3.1.8. Suppose that $\Omega \subseteq \Lambda$ is an L-invariant o-convex subgroup containing C. For an (L, C)-quasi-isometry $f : X \longrightarrow X'$, the restriction $L|_\Omega : \Omega \longrightarrow \Omega$ yields an (L, C)-quasi-isometry $X_o^\Omega \longrightarrow X'^\Omega_{f(o)}$ of Ω-metric spaces. Similarly, the corestriction $\bar{L} : \Lambda/\Omega \longrightarrow \Lambda/\Omega$ yields a bi-Lipschitz map (an $(\bar{L}, 0)$-quasi-isometry) $\bar{f} : X/\Omega \longrightarrow X'/\Omega$ of Λ/Ω-metric spaces.

While all these metric concepts are rather simple and basic, we will see in Section 3.3 that the construction of *asymptotic cones* is just a special instance of these basic methods, applied to ultrapowers of metric spaces.

3.2. Valuations on real closed fields

The basic reference for this section is again [24]; the ideas date back to Baer [4] and Artin–Schreier [2,3].

3.2.1. Let R be a *real closed field*, i.e. an ordered field where every positive element is a square, and where every odd polynomial has a zero. Let $O \subseteq R$ be an o-convex subring (containing 0 and 1). Then O is a *valuation ring*: if $\alpha \in R \setminus O$, then $\alpha^{-1} \in O$. The ring O thus has a unique maximal ideal M consisting of all nonunits. The quotient O/M is again a real closed field. If K is any maximal subfield of O, then K is real closed and o-projects onto O/M, so O splits (not naturally) as a direct sum $R = K \oplus M$,

$$0 \longrightarrow M \longrightarrow O \overset{\leftarrow}{\longrightarrow} O/M \longrightarrow 0.$$

An archimedean real closed field R (such as $\mathbb{R}$) contains no proper o-convex subring, but there is an abundance of nonarchimedean examples; in the next section, we will have a construction using ultraproducts.

3.2.2. Let R be a nonarchimedean real closed field, let $\alpha \gg 1$, and let $O^{\langle\alpha\rangle} = \{r \in R \mid |r| \leq \alpha^n \text{ for some } n \in \mathbb{N}\}$ denote the o-convex subring generated by α. The maximal ideal is $M^{\langle\alpha\rangle} = \{r \in R \mid |r| \leq \alpha^{-n} \text{ for all } n \in \mathbb{N}\}$, and we put

$$R^{\langle\alpha\rangle} = O^{\langle\alpha\rangle}/M^{\langle\alpha\rangle}.$$

Intuitively, we have truncated the field R behind the powers of α. If $R = R^{\langle\alpha\rangle}$, we call R an *α-archimedean field* (because R is archimedean over the subfield $\mathbb{Q}(\alpha)$). The real closed field $R^{\langle\alpha\rangle}$ is $\bar{\alpha}$-archimedean, where $\bar{\alpha}$ is the image of $\alpha \in O^{\langle\alpha\rangle}$ in $R^{\langle\alpha\rangle}$.

3.2.3. In general, a real closed field need not have a logarithm or exponential function (i.e. an o-isomorphism $(R, +) \overset{\cong}{\longrightarrow} (R_{>0}, \cdot)$). In fact, no α-archimedean real closed field can have an exponential function. However, we may define a *formal logarithm* as follows: we take for $(\Lambda, +)$ an o-isomorphic copy of the ordered abelian (multiplicative) group $(R_{>0}, \cdot)$ of positive elements. For this, we rewrite this group additively, putting

$$\lg : (R_{>0}, \cdot) \cong (\Lambda, +).$$

Note that then $|\lg r|$ corresponds to $\max\{r, r^{-1}\}$. If $O \subseteq R$ is an o-convex subring, we may consider the o-convex subgroup

$$\Omega = \{\lg r \mid r > 0 \text{ is a unit of } O\} \subseteq \Lambda.$$

If we put $\nu(0) = \infty$, then the map

$$\nu : r \longmapsto -\lg|r| + \Omega \in \Lambda/\Omega$$

is precisely the *valuation* determined by $O \subseteq R$; the *value group* is $\Gamma = \Lambda/\Omega$.

3.2.4. Suppose that $R = R^{\langle\alpha\rangle}$ is α-archimedean, and consider the o-convex subring

$$O_{\langle\alpha\rangle} = \{r \in R \mid |r|^n \leq \alpha \text{ for all } n \in \mathbb{N}\}.$$

Then the value group $\Gamma = \Lambda/\Omega$ is archimedean, because $\Lambda = \Lambda^{\langle\lg\alpha\rangle}$ and $\Omega = \Lambda_{\langle\lg\alpha\rangle}$, so we have a (nondiscrete, rank 1) real-valued o-valuation ν on the field R. A special instance of this construction is *Robinson's field* ${}^{\rho}\mathbb{R}$ [25] [19], which we introduce in the next section.

3.3. Ultraproducts and asymptotic cones

A basic reference for ultraproducts is [10].

3.3.1. Let I be a countably infinite set, and let $\mu : 2^I \longrightarrow \{0, 1\}$ be a finitely additive probability measure which is not a Dirac measure (i.e. $\mu(\text{finite}) = 0$). The collection of all sets with μ-measure 1 is a *nonprincipal ultrafilter*. Suppose that $(\mathcal{Q}_i)_{i\in I}$ is a family of first order structures: e.g. a family of groups, rings, fields, or metric spaces. In the direct product $\prod \mathcal{Q}_i$ of these structures, we identify two elements $(q_i)_{i\in I}, (q_i')_{i\in I}$ if the set $\{j \in I \mid q_j \neq q_j'\}$ has μ-measure 0. The resulting collection of equivalence classes is the *ultraproduct* $\prod_\mu \mathcal{Q}_i$; it has the same first-order properties as its factors and is $\aleph_1$-saturated.

An important special case is when all structures $\mathcal{Q}_i$ are equal to one fixed structure $\mathcal{Q}$; in this case, one obtains an *ultrapower*

$${}^*\mathcal{Q} = \prod_\mu \mathcal{Q}$$

which is also called a *nonstandard model* of $\mathcal{Q}$. If $\mathcal{Q} = (\mathbb{R}, +, \cdot, \leq)$ is the field of real numbers, the resulting field $\prod_\mu(\mathbb{R}, +, \cdot, \leq) = ({}^*\mathbb{R}, +, \cdot, \leq)$ is the field of *nonstandard real numbers*. This is a nonarchimedean real closed field.

3.3.2. Let $\alpha \in {}^*\mathbb{R}$ be an infinitely large nonstandard real, and let $\rho = \alpha^{-1}$. The field

$${}^{\rho}\mathbb{R} = ({}^*\mathbb{R})^{\langle\alpha\rangle}$$

was first studied by Robinson [25] [19]. We call it *Robinson's field.*

3.3.3. Suppose now that each $\mathcal{Q}_i = (d_i : X_i \times X_i \longrightarrow \mathbb{R})$ is an $\mathbb{R}$-metric space. Then the ultraproduct

$$\prod_\mu (d_i : X_i \times X_i \longrightarrow \mathbb{R}) = ({}^*d : {}^*X \times {}^*X \longrightarrow {}^*\mathbb{R})$$

is a Λ-metric space, with $\Lambda = ({}^*\mathbb{R}, +)$. If we pick a point $o \in {}^*X$, a number $\alpha \gg 1$ in ${}^*\mathbb{R}$, then

$$X_o^{\langle\alpha\rangle} = \{x \in^* X \mid {}^*d(x, o) \leq n\alpha \text{ for some } n \in \mathbb{N}\}$$

and the $\mathbb{R}$-metric space

$$Cone(X) = Cone(X, \mu, \alpha^{-1}, o) = {}^*X_o^{(\alpha)} = {}^*X_o^{\langle\alpha\rangle}/\Lambda_{\langle\alpha\rangle}$$

is precisely the *asymptotic cone* of (X, d), where all points x, y with infinitesimal distance ${}^*d(x, y)\frac{1}{\alpha}$ are identified [12]. This general construction is due to van den Dries and Wilkie [32]. It generalizes the notion of Gromov–Hausdorff convergence of metric spaces, enforcing convergence, see [8] I.5.51.

3.3.4. A variation of this is the *ultracone*

$$UCone(X) = {}^*X/\mathbb{R}_{\text{fin}},$$

where any two points in ${}^*X$ at *finite distance* are identified.

3.3.5. Let $f : X \longrightarrow X'$ be an (L, C)-quasi-isometry of $\mathbb{R}$-metric spaces. Its non-standard extension ${}^*f : {}^*X \longrightarrow {}^*X'$ is also an (L, C)-quasi-isometry (of ${}^*\mathbb{R}$-metric spaces), and it is immediate from 3.1.7 that f induces bi-Lipschitz homeomorphisms

$$Cone(X) \overset{Cone(f)}{\longrightarrow} Cone(X') \quad \text{and} \quad UCone(X) \overset{UCone(f)}{\longrightarrow} UCone(X'),$$

i.e. the asymptotic cone and the ultracone are functors on the category of (pointed) metric spaces and quasi-isometries, and quasi-isometric spaces have (bi-Lipschitz) homeomorphic asymptotic cones and ultracones.

The following lemma is a straight-forward consequence of $\aleph_1$-saturatedness (a strong kind of completeness of ultrapowers).

Lemma 3.3.6. *Two quasi-isometries $f, g : X \longrightarrow X'$ have bounded distance if and only if they have equal ultracones, $UCone(f) = UCone(g)$. The quasi-isometry group $\mathcal{QI}(X)$ acts as a group of bi-Lipschitz maps on $UCone(X)$.* □

3.3.7. The definition of the asymptotic cone depends on the ultrafilter μ. However, in many cases, this ultrafilter does not really matter.

(1) Let $\Gamma = \mathbb{Z}^n$ be the free abelian group in n-generators equipped with the word metric defined in 3.1.2 (which for the canonical generators coincides with the Manhattan Taxi metric, $d(x, y) = |x_1 - y_1| + \cdots + |x_n - y_n|$). It is easy to see that the asymptotic cone (with respect to the canonical basepoint) then is the topological group $\mathbb{R}^n$, also with the Manhattan Taxi metric. Hence it is independent of the ultrafilter.
(2) Let Γ be the free group of rank n with the word metric. Its asymptotic cone is a complete homogeneous $\mathbb{R}$-tree and again independent of the ultrafilter.
(3) Point showed in [23] that if Γ is a finitely generated nilpotent group, then the asymptotic cone is homeomorphic to $\mathbb{R}^m$, where m is the sum of the ranks of the descending central series of G.

3.3.8. Gromov asked the following question [12] p. 42: Is there a *finitely presented* group, for which different ultrafilters yield non-homeomorphic asymptotic cones?

Examples of *finitely generated* groups with non-homeomorphic asymptotic cones were given by Thomas and Velickovic [28]. We will answer this question assuming $(\neg\mathsf{CH})$ using the fact that uniform lattices in Lie groups are finitely presented groups (see Section 3.7).

3.4. Affine Λ-buildings and spherical buildings

Affine Λ-buildings should be viewed as higher-dimensional versions of trees or Λ-trees (for the latter, see [1]). They are patched together from copies of affine Λ-apartments in a similar way as a manifold or algebraic variety is patched together from coordinate charts.

Affine Λ-apartments are generalizations of affine reflection groups acting on Euclidean space; they are the natural coordinate charts for affine Λ-buildings, and they model the flats of Riemannian symmetric spaces over arbitrary real closed fields (see Section 5).

Basic references for root systems and Weyl groups are Bourbaki [7] and Onishchik–Vinberg [21, 4.§2].

3.4.1. Let Φ be a root system, with associated reflection group W (i.e. Φ is a finite set of vectors in some Euclidean space, satisfying certain axioms, and W is the group generated by the Euclidean reflections along these vectors). A typical example is the root system of type A_{m-1} consisting of all sums and $\pm e_i \pm e_j$, for $i \neq j$, where $e_1, \dots, e_m$ is the standard basis of $\mathbb{R}^m$. (More precisely, these vectors form a root system in the hyperplane they generate.) The group generated by the reflections along these roots is the symmetric group $W = \mathrm{Sym}(m)$, acting by coordinate permutations on $\mathbb{R}^m$. (In the examples we are interested in later, W will be the *Weyl group* $W = N/(B \cap N)$ of a semisimple Lie group G, see 3.4.4 below; the root system of type A_{m-1} belongs to the group $\mathrm{SL}_m\mathbb{R}$).

Then W acts naturally on the $\mathbb{Z}$-lattice $L \cong \mathbb{Z}^n$ spanned by the roots. For a divisible ordered abelian group Λ, we have a natural action of W on $L_\Lambda = L \otimes \Lambda \cong \Lambda^n$. Let $W_\Lambda = W \ltimes L_\Lambda$ denote the corresponding affine reflection group. Using the coroots (which are certain homomorphisms $L \longrightarrow \mathbb{Z}$) it is possible to construct a W_Λ-invariant Λ-metric $d : L_\Lambda \times L_\Lambda \longrightarrow \Lambda$ (explicitly, $d(x, y) = \sum_{\xi^\vee} |\xi^\vee(x - y)|$, the sum being taken over all coroots). For the case of the symmetric group, this metric is essentially the Manhattan Taxi Metric. Using the action of W_Λ on L_Λ, it makes sense to talk about *reflection hyperplanes* (the fixed point sets of reflections), *half-spaces* (the points x with $\xi^\vee(x) \geq 0$ for a fixed coroot), and *Weyl chambers* (the intersection of all halfspaces corresponding to a set of positive roots).

Finally, a subset $B \subseteq L_\Lambda$ is called *convex* if it is the intersection of finitely many affine halfspaces.

a reflection hyperplane

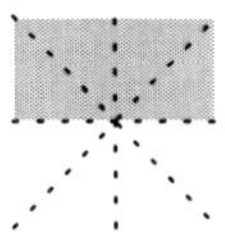
a half-space

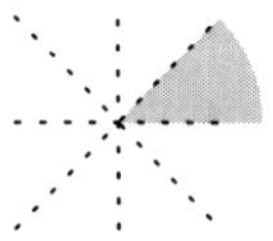
a Weyl chamber

3.4.2. Suppose that $\mathcal{I}$ is a Λ-metric space. An *atlas* on $\mathcal{I}$ is a collection $\mathcal{A}$ of Λ-isometric injections $\phi : L_\Lambda \longrightarrow \mathcal{I}$, called *coordinate charts*, with the following properties.

(A1) If ϕ is in $\mathcal{A}$ and $w \in \overline{W}$, then $\phi \circ w : L_\Lambda \longrightarrow \mathcal{I}$ is in $\mathcal{A}$.

(A2) Given two charts ϕ_1, ϕ_2, the set $B = \phi_1^{-1}(\phi_2(L_\Lambda))$ is convex, and there exists a $w \in \overline{W}$ with $\phi_1|_B = \phi_2 \circ w|_B$.

The sets $F = \phi(L_\Lambda) \subseteq \mathcal{I}$ are called *apartments*; the image $S = \phi(S_0)$ of a Weyl chamber $S_0 \subseteq L_\Lambda$ is called a *sector*.

(A3) Given $x, y \in \mathcal{I}$, there exists an apartment $F = \phi(L_\Lambda)$ containing x and y.

(A4) Given two sectors $S_1, S_2 \subseteq \mathcal{I}$, there exist subsectors $S_1' \subseteq S_1$ and $S_2' \subseteq S_2$ and an apartment F containing $S_1' \cup S_2'$.

(A5) If F_1, F_2, F_3 are apartments such that each of the three sets $F_i \cap F_j, i \neq j$, is a halfapartment (i.e. the ϕ-image of a halfspace), then $F_1 \cap F_2 \cap F_3 \neq \emptyset$.

(A6) For any apartment F and any $x \in F$, there exists a retraction $\rho_{x,F} : \mathcal{I} \longrightarrow F$ which diminishes distances, with $\rho_{x,F}^{-1}(x) = \{x\}$.

The pair $(\mathcal{I}, \mathcal{A})$ is called an *affine* Λ*-building* of rank n [5].

If we take for $\mathcal{I}$ a Riemannian symmetric space and the maximal flats as the apartments (see 3.5.2 below), then $\mathcal{I}$ satisfies axioms **(A1)** and **(A3)**, and, rather trivially, axioms **(A2)** and **(A5)**, while axiom **(A4)** is only approximately true.

For $n = 1$, an affine Λ-building is the same as a Λ-tree without dead ends; affine Λ-buildings are in a sense higher-dimensional versions of (Λ-) trees (see [1] for more information on Λ-trees). The notion of affine Λ-buildings is due to Bennett [5]; they are generalizations of Tits' *systèmes d'appartements* [31] (the case where $\Lambda = \mathbb{R}$).

3.4.3. Associated to a finite reflection group W is a certain simplicial complex $\Sigma(W)$, the *Coxeter complex*: roughly speaking, $\Sigma(W)$ is a W-invariant triangulation of the unit sphere in the Euclidean vector space on which W is acting naturally. A *spherical building* is an abstract simplicial complex (a poset) Δ with a distinguished collection of subcomplexes Σ, called *apartments*, satisfying certain axioms: each apartment $\Sigma \subseteq \Delta$ is isomorphic to a (fixed) Coxeter complex $\Sigma(W)$, any two simplices are contained in some apartment, and given two apartments $\Sigma_1, \Sigma_2 \subseteq \Delta$, there exists an isomorphism $\Sigma_1 \longrightarrow \Sigma_2$ fixing $\Sigma_1 \cap \Sigma_2$ simplex-wise. Note the analogy with the axioms **(A1)–(A3)** above. For more details, we refer to [9], [27], [30].

3.4.4. Let G be a noncompact semisimple real Lie group with Iwasawa decomposition $G = KAU$ (see [15] p. 317 — in the Lie group literature, the unipotent group U is denoted N, but this conflicts with the notation for BN-pairs). Then there is a natural BN-pair of G obtained by putting $N = N_G(A)$ and $B = K_0AU$ where $K_0 = Z_K(A)$. This BN-pair gives rise to a spherical building $\Delta(G, \mathbb{R})$ where the set of apartments corresponds to the set of cosets G/N, see [33] Ch. 1.2 and [6] §24.C. The groups G, B and N can be chosen to be real algebraic groups defined over $\mathbb{Q}$, and for any real closed field R, the groups $B(R)$ and $N(R)$ form the canonical BN-pair of the algebraic group $G(R)$. This spherical building can also be described as the poset of all cosets of all subgroups $P < G$ containing B, ordered by reversed inclusion (see [9] p. 111).

3.4.5. Let $\mathcal{I}$ be an affine Λ-building, let $o \in \mathcal{I}$. Consider the collection $\mathbf{Sec}_o$ of all sectors $\phi(S_0) \subseteq \mathcal{I}$ whose tip $\phi(0)$ is o (recall that $S_0 \subseteq L_\Lambda$ is a Weyl chamber). Then $\mathbf{Sec}_o$ generates in a rather natural way a poset, the *spherical building at infinity*

$\partial(\mathcal{I}, \mathcal{A})$. Using the axioms above, one can show that different base points lead to naturally isomorphic buildings at infinity.

Let R be a real closed field and let $O \subsetneq R$ be an o-convex subring. Let G be a semisimple real Lie group without compact factors, with Iwasawa decomposition $G = KAU$. Our first main result is as follows.

Theorem 3.4.6. *Let G be a semisimple Lie group without compact factors. The quotient $\mathcal{I} = G(R)/G(O)$ is an affine Γ-building, where $\Gamma \cong R^*/O^*$ is the value group of the valuation determined by O, see 3.2.3. In particular, if $R = R^{\langle a \rangle}$ is a-archimedean and $O = O_{\langle a \rangle}$, then $G(R)/G(O)$ is an affine $\mathbb{R}$-building.*

For the apartment system $\mathcal{A}$, we take the images of the maximal flats in the Riemannian symmetric space $G(R)/K(R)$ under the canonical map $G(R)/K(R) = G(R)/K(O) \longrightarrow G(R)/G(O)$, see Section 3.5. We denote this affine Γ-building by

$$\Delta(G, R, O) = (\mathcal{I}, \mathcal{A}).$$

The spherical building at infinity is the one coming from the natural BN-pair of $G(R)$, i.e. $\partial(\mathcal{I}, \mathcal{A}) = \Delta(G, R)$. If no factor of G has rank 1, the little projective group *of $\partial(\mathcal{I}, \mathcal{A})$ (the group generated by all root elations, which is normal in the automorphism group of $\mathcal{I}$) is precisely (the identity component of) the group $G(R)$.*

The proof depends on various classical results about semisimple Lie groups and on the fact that the theory of real closed fields is complete. Then axiom **(A3)** is a consequence of the KAK-decomposition, $G(R) = K(R)A(R)K(R)$, while **(A2)** follows from Kostant's Convexity Theorem [16]. The proof of **(A6)** depends also on the Convexity Theorem, and properties of the Iwasawa projection $g = kau \longmapsto a$.

3.5. Riemannian symmetric spaces over real closed fields

In this section we explain how to recover in a Riemannian symmetric space the metric and other geometric information from its algebraic presentation as a coset space of a real Lie group. This allows us to extend the concept of a Riemannian symmetric space to arbitrary real closed fields, which will be crucial in our approach. For the geometry of Riemannian symmetric spaces, see [8], [11], and [13].

3.5.1. A Riemannian symmetric space X is a Riemannian manifold for which the geodesic reflection at each point extends to a global isometry. The Riemannian structure yields a (complete) real-analytic metric $d_{\mathbb{R}}$ on X. Such a Riemannian symmetric space decomposes uniquely as a metric product of a *compact symmetric space*, a *Euclidean space*, and a *symmetric space of noncompact type*. To us, only symmetric spaces of noncompact type will be important (because all compact metric spaces are quasi-isometric to a point, and one cannot say much about quasi-isometries of Euclidean space), so we assume that X has no compact or Euclidean factors. Then X can be written as a homogeneous space $X = G/K$, where G is a centerless semisimple Lie group without compact factors, and $K \subseteq G$ a maximal compact subgroup.

3.5.2. By definition, a *flat* in such a Riemannian symmetric space $X = G/K$ is a subspace isometric to $\mathbb{R}^n$ for some n. The maximal n for which a flat exists is the rank of X (or G) and flats isometric to $\mathbb{R}^n$ are called *maximal flats*. These maximal flats can be described in the group using the canonical BN-pair of G as in 3.4.4. It can be shown that for any maximal flat $F \subseteq X$ there is some $h \in G$ such that $F = \{gK : gK \cap hN \neq \emptyset\}$.

We say that X has a *factor of rank* 1 if the semisimple group G has a simple factor of rank 1. Geometrically, a rank 1 factor is a direct metric factor isometric to hyperbolic space over the reals, the complex numbers, the quaternions, or the Cayley numbers.

Similarly as an affine Λ-building, a Riemannian symmetric space has a *spherical building at infinity*, which is usually described in terms of geodesic rays emanating from some point o (see [8] p. 337) and the combinatorics of the maximal flats passing through o. This building is naturally isomorphic to the spherical building $\Delta(G, \mathbb{R})$ introduced in 3.4.4.

As mentioned in Section 3.4 the collection of maximal flats of X satisfies some, but not all of the axioms of an affine Λ-building. In the asymptotic cone these flats turn into the apartments of the corresponding Λ-building. This explains why the building at infinity of the affine Λ-building we obtain is of the same type as the building at infinity of the Riemannian symmetric space X.

3.5.3. We consider the example $G = \mathrm{SL}_n\mathbb{R}$ with its maximal compact subgroup $K = \mathrm{SO}(n)$. Let P_n denote the set of all symmetric positive definite real $n \times n$-matrices with determinant 1. Every such matrix is of the form $X = gg^T$, for some $g \in \mathrm{SL}_n\mathbb{R}$. The set P_n can be identified with the Riemannian symmetric space $SL_n\mathbb{R}/\mathrm{SO}(n)$. The group $\mathrm{SL}_n\mathbb{R}$ acts as

$$(g, X) \longmapsto gXg^T$$

(the geodesic reflection at $X \in P_n$ is $Y \longmapsto XY^{-1}X$). Up to a real scaling factor, the analytic metric d_R induced by the Riemannian metric of P_n is given as follows. For $X \in P_n$, let $Y = \log(X)$ denote the unique traceless symmetric matrix with $\exp(Y) = X$. Then $d_R(\mathbf{1}, X)^2 = \mathrm{tr}(Y^2)$.

3.5.4. By definition a Riemannian symmetric space is a smooth manifold. But since semisimple Lie groups and their maximal compact subgroups are real algebraic groups, it makes sense to view Riemannian symmetric spaces as algebraic varieties, which we may consider over any real closed field. Thus we view them as a coset space $G(R)/K(R)$, for any real closed field R, and semisimple real algebraic group G, with a reductive subgroup K such that $K(\mathbb{R})$ is a maximal compact subgroup of $G(\mathbb{R})$. We may now transfer the concept of maximal flats to this setting using the canonical BN-pair for $G(R)$ and the algebraic description in 3.5.2.

3.5.5. We now describe what kind of metric we will impose on these spaces. Since P_n is an algebraic variety defined over $\mathbb{Q}$, we may consider its set of R-points $P_n(R) \cong \mathrm{SL}_n(R)/\mathrm{SO}(n, R)$. over any real closed field R. The problem is that the real analytic metric d_R need not be defined over an arbitrary real closed field R. We fix this problem, using the formal logarithm $\lg : R_{>0} \longrightarrow \Lambda$ defined in 3.2.3. Given a diagonal matrix $D = \mathrm{diag}(x_1, \dots, x_n) \in P_n(R)$, we define the Λ-*valued distance*

$$d(\mathbf{1}, D) = |\lg x_1| + \cdots + |\lg x_n| \in \Lambda.$$

This function is obviously invariant under coordinate permutations (the Weyl group action for $\mathrm{SL}_n R$); it follows that we can use the $\mathrm{SL}_n R$-action to extend d to a well-defined distance function on $P_n(R)$, setting

$$d(gg^T, gDg^T) = d(\mathbf{1}, D)$$

(where D is a diagonal matrix). On the set F of all diagonal matrices in $P_n(R)$, this is clearly a Λ-metric (essentially, it is the Manhattan Taxi Metric). Using Kostant's Convexity Theorem, on sees that d is indeed an $\mathrm{SL}_n R$-invariant metric on $P_n(R)$. Over the reals, we may take $\Lambda = \mathbb{R}$ and $\lg = \log$; then d and d_R are different, but (bi-Lipschitz) equivalent metrics.

3.5.6. We have successfully made the symmetric space $X(R) = P_n(R)$ into a Λ-metric space. Suppose now that $O \subseteq R$ is an o-convex subring. Put as before $\Omega = \{\lg r \mid r > 0 \text{ and } r \text{ is a unit of } O\} \subseteq \Lambda$. Then $\Gamma = \Lambda/\Omega$ is the value group of the valuation determined by O. The group $\mathrm{SL}_n R$ acts by isometries on the Λ/Ω-metric space $P_n(R)/\Omega$, and it is not difficult to check the following result:

$$P_n(R)/\Omega = \mathrm{SL}_n(R)/\mathrm{SL}_n(O).$$

3.5.7. Suppose now that $R = R^{\langle\alpha\rangle}$ is α-archimedean, see 3.2.2, and that $O = O_{\langle\alpha\rangle}$. Then the value group $\Gamma = \Lambda/\Omega = \Lambda^{(\alpha)}$ is archimedean (we could take for example $R = {}^{\rho}\mathbb{R}$, Robinson's field). Consequently,

$$P_n(R)/\Omega = \mathrm{SL}_n(R)/\mathrm{SL}_n(O)$$

is an $\mathbb{R}$-metric space. The main result of Section 3.4 is that this quotient is always an (nondiscrete) affine Λ-building (of type $\widetilde{\mathsf{A}}_{n-1}$, in fact). The group $\mathrm{SL}_n(R)$ acts as a transitive automorphism group on its vertices.

3.5.8. We have concentrated on the Riemannian symmetric space $P_n = \mathrm{SL}_n\mathbb{R}/\mathrm{SO}(n)$. Everything we have done works in the same generality for any Riemannian symmetric space of noncompact type. In fact, any irreducible Riemannian symmetric space of noncompact type X can be embedded into P_n in such a way that the analytic metric d_R of X is (up to rescaling) the metric inherited from the one on P_n; the embedding can be chosen to be algebraic over $\mathbb{Q}$. Then one obtains without much difficulty the following result.

Theorem 3.5.9. *Let R be a (nonarchimedean) real closed field, let O be an o-convex subring, let X be a Riemannian symmetric space of noncompact type. Let G be the corresponding semisimple real Lie group, with maximal compact subgroup K. Then we can view G and K as real algebraic groups defined over $\mathbb{Q}$, and $X = G/K$ as a real algebraic variety. Define Λ and Ω as above. Then there exists a $G(R)$-invariant Λ-metric on $X(R) = G(R)/K(R)$ (semi-algebraic over $\mathbb{Q}$), and there is a natural equivariant identification*

$$X(R)/\Omega = G(R)/G(O)$$

of Λ/Ω-metric spaces. If R is in addition α-archimedean, and $O = O_{\langle\alpha\rangle}$, then $X(R)/\Omega$ is an $\mathbb{R}$-metric space. □

Now we consider as a special case the asymptotic cone. Let ${}^*X$ be the ultrapower of X, pick a basepoint $o \in {}^*X$ and an infinitely large nonstandard real $\alpha \gg 1$. Let $\rho = 1/\alpha$.

Corollary 3.5.10. *The asymptotic cone of the Riemannian symmetric space X is*

$$Cone(X) = G(R)/G(O),$$

where $R = {}^\rho\mathbb{R} = ({}^*\mathbb{R})^{\langle\alpha\rangle}$ is Robinson's asymptotic field, and $O = O_{\langle\bar{\alpha}\rangle}$. It can be identified with the affine $\mathbb{R}$-building $\Delta(G, {}^\rho\mathbb{R}, O)$. The ultracone of X is

$$UCone(X) = G({}^*\mathbb{R})/G({}^*\mathbb{R}_{\text{fin}});$$

it can be identified with the affine $\mathbb{R}$-building $\Delta(G, {}^*\mathbb{R}, {}^*\mathbb{R}_{\text{fin}})$. □

The fact that as a metric space, $Cone(X) = G({}^\rho\mathbb{R})/G(O)$ was proved independently in [29].

Affine $\mathbb{R}$-buildings are rigid in the sense that any homeomorphism preserves the building structure. This important fact was first proved in [14].

Proposition 3.5.11 ([14,17]). *Let $(\mathcal{I}, \mathcal{A})$ and $(\mathcal{I}', \mathcal{A}')$ be metrically complete affine $\mathbb{R}$-buildings with maximal atlases $\mathcal{A}$ and $\mathcal{A}'$, respectively. Then any homeomorphism of $\mathcal{I}$ onto $\mathcal{I}'$ preserves apartments; moreover, it induces an isomorphism $\partial(\mathcal{I}, \mathcal{A}) \cong \partial(\mathcal{I}', \mathcal{A}')$.* □

The proof in [17] uses the fact that each apartment yields a global section in the sheaf of local homology groups of $\mathcal{I}$, and that the resulting sections can be characterized by their local properties.

3.6. Proof of the Margulis conjecture

We can now outline a proof of the Margulis conjecture, using the previous results. Let X and X' be Riemannian symmetric spaces of noncompact type, and let

$$f : X \longrightarrow X'$$

be an (L, C)-quasi-isometry. Taking an ultrapower of (X, X', f), we obtain an (L, C)-quasi-isometry

$${}^*f : {}^*X \longrightarrow {}^*X'$$

between the ${}^*\mathbb{R}$-metric spaces ${}^*X, {}^*X'$. We use the metric d introduced in 3.5.5, which is bi-Lipschitz equivalent to the analytic metric d_R induced by the Riemannian metric.

3.6.1. As we remarked in 3.3.5, we obtain bi-Lipschitz maps

$$Cone(X) \overset{Cone(f)}{\longrightarrow} Cone(X') \quad \text{and} \quad UCone(X) \overset{UCone(f)}{\longrightarrow} UCone(X').$$

For the rank 1 factors, we note the following. The asymptotic cone of a hyperbolic space is a homogeneous $\mathbb{R}$-tree (its ultracone is a homogeneous ${}^*\mathbb{R}/{}^*\mathbb{R}_{\mathrm{fin}}$-tree). By 3.4.6 and 3.5.11 we can already conclude that the groups G and G' have the same number of rank 1 factors, and that their simple factors of higher rank are isomorphic. If X has no rank 1 factors, then the same is true for X', and X and X' are isometric (after rescaling the metrics on the factors of X', say). This gives already the first half of the Margulis conjecture:

Proposition 3.6.2. *Let X, X' be Riemannian symmetric spaces of noncompact type. If X and X' are quasi-isometric, then X and X' have the same number of rank 1 factors; their higher rank factors are isometric (possibly after rescaling the metrics on the factors).* □

3.6.3. In order to find an isometry $X \longrightarrow X'$ at bounded distance from f, we argue as follows: By repeated applications of compactness arguments (and using that ultrapowers are $\aleph_1$-saturated) we see from the fact that $Cone(f)$ takes apartments to apartments in each asymptotic cone (i.e. for any choice of the base point o and the scaling factor ε) that there is some bound $D \in {}^*\mathbb{R}$ such that for any flat F of ${}^*X$ is there is a flat F' of ${}^*X'$ such that the image of F under ${}^*f$ has Hausdorff distance at most D from F'. Note that this is clearly first order expressible and so this holds also in $f : X \longrightarrow X'$; in particular, the bound D may be chosen to be a finite number. Thus, the ultracone $UCone(f)$ of f preserves the apartments of the affine ${}^*\mathbb{R}/{}^*\mathbb{R}_{\mathrm{fin}}$-building $\Delta(G, {}^*\mathbb{R}, {}^*\mathbb{R}_{\mathrm{fin}})$. This implies that f induces a definable isomorphism $\partial UCone(f) : \Delta(G, {}^*\mathbb{R}) \longrightarrow \Delta(G', {}^*\mathbb{R})$ of the respective spherical buildings at infinity. A simple nonstandard argument shows that this map is continuous; we obtain thus a continuous definable isomorphism $g : \Delta(G, \mathbb{R}) \longrightarrow \Delta(G', \mathbb{R})$ for the spherical buildings of the Lie groups G, G'.

Up to this point, our reasoning is valid without any assumptions on the ranks of the simple factor of G and G'. To get further, we use the following result.

Proposition 3.6.4. *Let G be a simple noncompact Lie group of rank at least 2. Then any continuous automorphism of the spherical building $\Delta(G, \mathbb{R})$ is real-algebraic and induced by an isometry of the corresponding symmetric space.* □

This is true, since the little projective group of $\Delta(G, \mathbb{R})$ is precisely (the connected component of) G. The little projective group is normal in the full automorphism group of $\Delta(G, \mathbb{R})$, and so the building automorphism induces a continuous automorphism of the Lie group G, which therefore has to be real algebraic. Note that the result is false in the rank 1 case, where the building $\Delta(G, \mathbb{R})$ is a just a topological space (a topological sphere) without further structure.

3.6.5. To finish the proof of the Margulis conjecture, we assume now that G has no simple factor of rank 1. By the above remarks we can find an isometry $g : X \longrightarrow X'$ which agrees with f on $\Delta(G, \mathbb{R})$ and thus on $\Delta(G, {}^*\mathbb{R})$. Since the latter building is determined by the apartments in the affine Λ-building $\Delta(G, {}^*\mathbb{R}, {}^*\mathbb{R}_{\mathrm{fin}})$, this implies that $UCone(g)(F) = UCone(f)(F)$ for any apartment F in $\Delta(G, {}^*\mathbb{R}, {}^*\mathbb{R}_{\mathrm{fin}})$. It follows that

$$UCone(f) = UCone(g),$$

and this completes the proof of the Margulis conjecture.

3.7. Finitely presented groups and a question of Gromov

We now turn to Gromov's question whether there exists a finitely presented group Γ with non-homeomorphic asymptotic cones. We use the following characterization of finitely presented groups (see [8] I.8.11):

Theorem 3.7.1. *A group Γ is finitely presented if and only if it acts properly and cocompactly by isometries on a 1-connected geodesic space X.*

Here, an action is called *cocompact* if there exists a compact set $C \subset X$ such that $X = \Gamma.C$. The action is called *proper* if there is some number $r > 0$ such that for each $x \in X$ the set $\{\gamma \in \Gamma \mid \gamma(B(x,r)) \cap B(x,r) \neq \emptyset\}$ is finite.

Riemannian symmetric spaces of noncompact type are in particular 1-connected and geodesic, and for these spaces, discrete subgroups of the isometry group which act cocompactly are called *uniform lattices*. Any Riemannian symmetric space of noncompact type allows many such uniform lattices. Note that such a lattice is quasi-isometric to the Riemannian symmetric space on which it acts and hence its asymptotic cone is homeomorphic to that of the symmetric space by the Milnor-Svarc-Lemma, see [8] I.8.16. Thus, if we can produce non-homeomorphic asymptotic cones of Riemannian symmetric spaces, we have also produced examples of finitely presented groups with non-homeomorphic asymptotic cones. The following is the crucial observation. It is based on 3.5.11, the $\aleph_1$-saturatedness of ultrapowers, and the fact the field of definition is definable in the building, provided that the rank is at least 2.

Proposition 3.7.2. *Let G and G' be noncompact and noncomplex simple Lie groups. The affine $\mathbb{R}$-buildings $G(R)/G(O)$ and $G'(R')/G'(O')$ obtained in 3.5.10 are homeomorphic if and only if $G \cong G'$ and $R \cong R'$ as valued fields (with valuation rings O and O'), respectively.*

Since we know that the asymptotic cone of the Riemannian symmetric space X is the affine $\mathbb{R}$-building

$$Cone(X) = G(R)/G(O),$$

where R is the Robinson field ${}^{\rho}\mathbb{R} = ({}^{*}\mathbb{R})^{\langle a \rangle}$ and $O = O_{\langle \bar{a} \rangle}$, it thus suffices to produce ultrafilters giving rise to non-isomorphic Robinson fields. It is not hard to see that if we assume (CH), then all Robinson fields are isomorphic since then any countable ultrapower is saturated. Hence in this case we will not be able to produce non-homeomorphic asymptotic cones from finitely presented groups acting on Riemannian symmetric spaces.

3.7.3. So we now assume ($\neg$CH). Examples constructed by Roitman [26] yield non-isomorphic ultrapowers differing from each other through their degree of saturation and the existence of scales. However, in the Robinson field we truncate behind some large nonstandard number α and divide out by the 'infinitesimal' part, which is exactly the part which contains the saturation.

Hence, in order to produce non-isomorphic asymptotic cones we concentrate on producing ultrafilters which give rise to ultrapowers of the natural numbers which differ in the coinitial segments. It is not hard to see that this suffices in order to make sure that the Robinson fields obtained from these ultrafilters will be non-isomorphic.

Thus, assuming $(\neg\mathsf{CH})$, the following result shows that there are finitely presented groups with non-homeomorphic asymptotic cones:

Theorem 3.7.4. *Assume* $(\neg\mathsf{CH})$. *Then there are uncountably many non-isomorphic Robinson fields.*

Corollary 3.7.5. *Let* Γ *be a finitely presented group quasi-isometric to a Riemannian symmetric space* $X = G/K$ *of noncompact type, and assume that* G *has at least one noncompact noncomplex simple factor of rank at least* 2. *Then* Γ *admits (uncountably many) non-homeomorphic asymptotic cones if and only if* $(\neg\mathsf{CH})$ *holds.*

Note that the Lie group G itself, endowed with any left-invariant Riemannian metric, is quasi-isometric to G/K (because the K-cosets in G have bounded diameter). Thus, our results apply also to the group G in place of G/K.

References

[1] R. Alperin and H. Bass, Length functions of group actions on Λ-trees, Combinatorial group theory and topology (Alta, Utah, 1984), 265–378, Princeton Univ. Press, Princeton, NJ (1987).

[2] E. Artin and O. Schreier, Algebraische Konstruktion reeller Körper, *Abh. Math. Sem. Univ. Hamburg*, **5** (1927) 85–99.

[3] E. Artin and O. Schreier, Eine Kennzeichnung der reell abgeschlossenen Körper, *Abh. Math. Sem. Univ. Hamburg*, **5** (1927) 225–231.

[4] R. Baer, Über nicht-Archimedisch geordnete Körper, Sitzungsber, *Heidelb. Akad. Wiss. Math.-Natur. Kl.*, **8** (1927) 3–13.

[5] C. Bennett, Affine Λ-buildings, *I. Proc. London Math. Soc.*, **68** (1994) 541–576.

[6] A. Borel, Linear algebraic groups, 2nd ed., Springer-Verlag, New York, Berlin, Heidelberg (1991).

[7] N. Bourbaki, Groupes et algèbres de Lie, ch. IV–VI, Paris, Hermann (1968).

[8] M. Bridson and A. Haefliger, Metric spaces of non-positive curvature, Springer-Verlag, Berlin (1999).

[9] K. Brown, Buildings, Springer-Verlag, New York (1989).

[10] C.C. Chang, H.J. Keisler, Model theory. 3rd ed. North-Holland Publishing Co., Amsterdam (1990).

[11] P. Eberlein, Geometry of nonpositively curved manifolds, University of Chicago Press, Chicago, IL, (1996).

[12] M. Gromov, Asymptotic invariants of infinite groups, Geometric group theory, Vol. 2 (Sussex, 1991), 1–295, Cambridge Univ. Press, Cambridge (1993).

[13] S. Helgason, Differential geometry, Lie groups, and symmetric spaces, Academic Press, Inc., New York, London (1978).

[14] B. Kleiner and B. Leeb, Rigidity of quasi-isometries for symmetric spaces and Euclidean buildings, *Inst. Hautes Études Sci. Publ. Math.*, (1997) No. 86, 115–197.

[15] A. W. Knapp, Lie groups beyond an introduction. Progress in Mathematics, 140. Birkhäuser Boston, Inc., Boston, MA (1996).

[16] B. Kostant, On convexity, the Weyl group and the Iwasawa decomposition, *Ann. Sci. École Norm. Sup. (4)*, **6** (1973) 413–455.

[17] L. Kramer and K. Tent, Asymptotic cones and ultrapowers of Lie groups, *Bull. Symbolic Logic*, **10** (2004) No. 2, 175–185.

[18] L. Kramer, S. Shelah, K. Tent and S. Thomas, Asymptotic cones of finitely presented groups, *Adv. Math.*, **193** (2005) no. 1, 142–173.
[19] W. Luxemburg, On a class of valuation fields introduced by A. Robinson, *Israel J. Math.*, **25** (1976) 189–201.
[20] G. D. Mostow, Strong rigidity of locally symmetric spaces, Princeton University Press, Princeton, N.J. (1973).
[21] A. L. Onishchik and E. B. Vinberg, Lie groups and algebraic groups, Springer-Verlag, Berlin (1990).
[22] A. Parreau, Dégénérescences de sous-groupes discrets de groupes de Lie semisimples et actions de groupes sur des immeubles affines, Thèse, Université de Paris-Sud XI Orsay (2000).
[23] F. Point, Groups of polynomial growth and their associated metric spaces, *J. Algebra*, **175** (1995) no. 1, 105–121.
[24] S. Prieß-Crampe, Angeordnete Strukturen: Gruppen, Körper, projektive Ebenen, Springer-Verlag, Berlin (1983).
[25] A. H. Lightstone and A. Robinson, Nonarchimedean fields and asymptotic expansions, North-Holland Publishing Co., New York (1975).
[26] J. Roitman, Nonisomorphic hyper-real fields from nonisomorphic ultrapowers, *Math. Z.*, **181** (1982) no. 1, 93–96.
[27] M. Ronan, Lectures on buildings, Academic Press, Inc., Boston, MA (1989).
[28] S. Thomas, B. Velickovic, Asymptotic cones of finitely generated groups, *Bull. London Math. Soc.*, **32** (2000) no. 2, 203–208.
[29] B. Thornton, Asymptotic cones of symmetric spaces, PhD Thesis, Univ. Utah (2002).
[30] J. Tits, Buildings of spherical type and finite BN-pairs, Lecture Note in Math. 386, Springer-Verlag, Heidelberg Berlin New York (1974).
[31] J. Tits, Immeubles de type affine, Buildings and the geometry of diagrams (Como, 1984), 159–190, Lecture Notes in Math. 1181, Springer-Verlag, Berlin (1986).
[32] L. van den Dries, A. Wilkie, Gromov's theorem on groups of polynomial growth and elementary logic, *J. Algebra*, **89** (1984) 349–374.
[33] G. Warner, Harmonic analysis on semi-simple Lie groups I, Springer-Verlag, Berlin Heidelberg New York (1972).

Proc. Inst. Wrk.shps. Int. Conf. – Essays in Geometric Group Theory No. 9, 2009, pp. 79–104.

Commensurators of Some Non-uniform Tree Lattices and Moufang Twin Trees

Peter Abramenko[1] and Bertrand Rémy[2]

[1]Department of Mathematics, Institut Camille Jordan – UMR CNRS, University of Virginia, Université Claude Bernard – Lyon 1, P.O. Box 400137, Charlottesville
[2]21 rue Claude Bernard, VA 22904 4137 – USA, 69622 Villeurbanne cedex, France
e-mail: pa8e@virginia.edu; remy@math.univ-lyon1.fr

Abstract. Sh. Mozes showed that the commensurator of the lattice $\mathrm{PSL}_2(\mathbf{F}_p[t^{-1}])$ is dense in the full automorphism group of the Bruhat–Tits tree of valency $p+1$, the latter group being much bigger than $\mathrm{PSL}_2(\mathbf{F}_p((t)))$. By G.A. Margulis's criterion, this density is a generalized arithmeticity result. We show that the density of the commensurator holds for many tree-lattices among those called of Nagao type by H. Bass and A. Lubotzky. The result covers many lattices obtained via Moufang twin trees.

Keywords. Tree, lattice of Nagao type, commensurator, twinning, Moufang property Kac-Moody group.

Subject Classification: 22F50, 22E20, 51E24, 22E40

Introduction

Tree lattices have been and still are the subject of a lot of interesting mathematical research, see in particular the monograph [BL01] and the papers cited there. One of the important issues is to find analogies – or to establish the differences – between lattices in the automorphism groups of locally finite trees and lattices in semisimple Lie groups (i.e. in semisimple algebraic groups over local fields). These questions are particularly relevant when the Lie groups are of rank 1. A fundamental theorem due to G.A. Margulis characterizes arithmetic lattices among lattices in semisimple Lie groups as those which are of infinite index in their commensurators [Mar91, Chapter IX, Theorem B]. (G.A. Margulis proved this characterization for finitely generated lattices, which excluded non cocompact lattices in rank 1 semisimple Lie groups over local fields with positive characteristic. A proof for this remaining case was later given by L. Lifschitz [Lif03].) G.A. Margulis also showed that being of infinite index in its commensurator already means that the commensurator of the lattice is ⟨⟨essentially⟩⟩ dense in the semisimple Lie group, and is in fact dense if the Lie group is simply connected (for a precise statement, see [Mar91, Chapter IX, Lemma 2.7]).

These results in the Lie group case motivated (among other things) the study of commensurators of tree lattices. In order to fix the ideas, we shall introduce some notation now. Let T be a locally finite tree with vertex set VT, and let $G := \mathrm{Aut}(T)$

be its group of automorphisms. Provided with the usual topology (a basis of open neighborhoods of the identity are the fixators of finite subtrees of T), the group G is locally compact. A T-*lattice* is by definition a lattice in G, i.e. a discrete subgroup of G of finite covolume. It is standard knowledge [BL01, Sections 1.5 and 3.2] that a subgroup Γ of G is a lattice in G if and only if all stabilizers Γ_x, $x \in VT$, are finite, as well as the sum $\mathrm{Vol}(\Gamma\backslash\backslash T) := \sum_{x \in \Gamma\backslash VT} |\Gamma_x|^{-1}$ is finite. We call Γ a *uniform* (resp. *non-uniform*) T-lattice if the quotient $\Gamma\backslash T$ is finite (resp. infinite). The *commensurator* of Γ in G is the group defined as: $\mathrm{Comm}_G(\Gamma) := \{g \in G | \Gamma \cap g\Gamma g^{-1}$ has finite index in Γ and in $g\Gamma g^{-1}\}$ and will be abbreviated by $C(\Gamma)$ in the following. Interesting questions concerning $C(\Gamma)$ are the following:

Question 1. Is Γ of infinite index in $C(\Gamma)$?

Question 2. Is $C(\Gamma)$ (essentially) dense in G?

As mentioned above, these two questions are equivalent for lattices in semisimple Lie groups, but it is known that they are not equivalent for tree lattices [BL01, Section 10.3]. So it is not quite clear which of the two conditions should be used to define, by analogy, ⟨⟨arithmetic⟩⟩ tree lattices. Usually one chooses the stronger condition and considers a tree lattice Γ as *arithmetic* if $C(\Gamma)$ is dense in G. An important theorem proved by Y. Liu states that all uniform tree lattices are arithmetic in the latter sense [Liu94]. Much less is known about commensurators of non-uniform tree lattices. In particular, to the best of our knowledge, only two examples of non-uniform tree lattices with dense commensurators are discussed in the literature, namely the example given in [BM96, Section 8.3] and the Nagao lattice $\mathrm{PSL}_2(\mathbf{F}_p[t^{-1}])$ which is shown in [Moz99] to have a dense commensurator for any prime number p.

It is the main objective of the present paper to generalize the two last mentioned examples in two directions. Firstly, both examples are *lattices of Nagao type* in the sense of [BL01, Chapter 10]. For a tree lattice Γ of Nagao type, we have a natural *level function* $\ell : VT \to \mathbf{N}$ which is Γ-invariant (see Definition 11 below), and we set $L := \{g \in G | \ell(g \cdot x) = \ell(x)$ for all $x \in VT\}$. Now the following question is crucial with respect to the examples discussed in [BM96] and [Moz99].

Question 3. Is $C(\Gamma) \cap L$ dense in L?

In Section 4.2 of this paper, we shall give a positive answer to this question for all lattices of *directly split Nagao type*, a class of lattices which we shall introduce in Section 4.1 below. This is our:

Theorem 4. *If Γ is a tree lattice of directly split Nagao type, then $C(\Gamma) \cap L$ is dense in L.*

Referring the reader to Section 4.1 for some technical details concerning lattices of directly split Nagao type, we just mention that this class of lattices is much larger than the examples discussed in [BM96] and [Moz99]. In particular, we allow arbitrary finite ⟨⟨root groups⟩⟩ whereas the root groups in [loc. cit.] are always cyclic. (For the classical Nagao lattice $\mathrm{PSL}_2(\mathbf{F}_q[t^{-1}])$ the root groups are isomorphic to the additive group of $\mathbf{F}_q$, hence cyclic if and only if q is a prime number.)

Now if the tree T is not biregular, then one easily checks that $L = G$ (Lemma 16), and hence a positive answer to Question 3 already means that $C(\Gamma)$ is dense in G. If, however, T is biregular (as in the case of the classical Nagao lattice), then we need a second ingredient in order to deduce the density of $C(\Gamma)$ from Theorem 4. Generalizing a strategy used by Sh. Mozes in [Moz99], we can show that $C(\Gamma)$ is already essentially dense in G in the biregular case if it is not contained in L. More precisely, denoting by G° the subgroup (of index at most 2) of G of all type–preserving automorphism of T (preserving the 2–colouring of T), we obtain:

Theorem 5. *If Γ is a T–lattice of directly split Nagao type with biregular T and $C(\Gamma)$ is not contained in L, then the closure of $C(\Gamma)$ in G contains G°.*

In general it is of course difficult to decide whether $C(\Gamma)$ is included in L or not. In the case of $\Gamma = \mathrm{PSL}_2(\mathbf{F}_q[t^{-1}])$, the additional ingredient that Sh. Mozes used was the transitive action of $C(\Gamma)$ on the set of edges of T. This transitivity property follows here, for the truly arithmetic group Γ, from the fact that its commensurator contains $\mathrm{PSL}_2(\mathbf{F}_q(t))$, an argument which is not available for other lattices of Nagao type. This brings us to our second line of generalization.

A special class of lattices of Nagao type arises from the theory of twin trees as follows. Let a locally finite (thick) twin tree $(T_\pm, \delta^*)$ be given which has the Moufang property (for definitions, see Section 4.3 below). We remark that the trees T_+ and T_- are necessarily biregular. Denote by A the automorphism group of the twin tree $(T_\pm, \delta^*)$ and by Λ its subgroup generated by all root groups. Fix a vertex of T_- and denote by Γ its stabilizer in Λ. Then it is well-known that Γ acts on T_+ as a non-uniform T_+-lattice with a ray R as fundamental domain. Moreover, Γ is always a lattice of Nagao type in the sense of [BL01], and it is of directly split Nagao type in our sense if it satisfies the following condition (which will also be explained in more detail in Section 4.3).

Assumption 6 (Comm). *For any prenilpotent pair $\{a; b\}$ of twin roots of the Moufang twin tree $(T_\pm, \delta^*)$, the corresponding root groups U_a and U_b commute.*

It is not difficult to check that A is contained in the commensurator of the T_+-lattice Γ and that it acts transitively on the set of all (geometric) edges of T. In particular, $C(\Gamma)$ is not contained in L, and thus Theorem 5 implies (where G° denotes the group of all type–preserving automorphisms of T_+):

Theorem 7. *If the thick locally finite Moufang twin tree $(T_\pm, \delta^*)$ satisfies* (Comm) *and Γ is the T_+-lattice described above, then the closure of $C(\Gamma)$ contains G°. Therefore $C(\Gamma)$ is dense in G if T_+ is not regular and also, in the regular case, if A contains an automorphism interchanging the types of vertices.*

First examples for groups Λ and Γ satisfying (Comm) come from rank 2 Kac-Moody groups over finite fields (see 4.3.4). As a special (affine) case we obtain $\Lambda = \mathrm{PSL}_2(\mathbf{F}_q[t, t^{-1}])$ and $\Gamma = \mathrm{PSL}_2(\mathbf{F}_q[t^{-1}])$ for any prime power q. In the latter case, the matrix $\left(\begin{smallmatrix} t & 0 \\ 0 & 1 \end{smallmatrix}\right) \in \mathrm{PGL}_2(\mathbf{F}_q[t])$ provides an element of A interchanging the vertex types: it acts on the tree as a hyperbolic translation, and its translation length is the length of a single edge. The construction of more exotic examples is sketched in

Example 69. As a very special case, we consider trees of constant degree equal to 7. For a non-uniform tree-lattice Γ naturally defined in terms of twinnings, the density of the commensurator holds, at least in the index 2 subgroup G°. Note that in this case the full automorphism group G is ⟨⟨as big as⟩⟩ in the case considered by Sh. Mozes, in the sense that the tree of valency 7 is ⟨⟨as homogeneous as⟩⟩ a Bruhat-Tits tree of some $PSL_2(\mathbf{F}_p((t)))$. Nevertheless, the case is new at least because it can easily be proved that the lattice itself cannot be linear over a field.

We close this introduction by an open question relevant to a higher-dimensional generalization of the above results.

Question 8. Let $\Delta_\pm$ be a locally finite thick Moufang twin building whose apartments are right-angled tilings of the hyperbolic plane. Does the non-uniform lattice naturally defined as a negative chamber stabilizer have a dense commensurator in $\mathrm{Aut}(\Delta_+)$?

We focus on the case of right-angled Fuchsian buildings because the local product structure of the links at vertices implies the existence of many symmetries for this class of buildings. Therefore the corresponding full automorphism groups are big topological groups. Recent results by F. Haglund on commensurators of some uniform lattices for hyperbolic buildings [Hag03] are analogous to Y. Liu's theorem for trees (and might be used as such). Note that concrete examples of right-angled Fuchsian buildings admitting a twinning are available [RR06, Theorem 4.E.2].

This paper is organized as follows. Section 4.1 deals with lattices of Nagao type as introduced by H. Bass and A. Lubotzky. The subclass of tree-lattices of directly split Nagao type is introduced (4.1.2) and the condition defining it is characterized in terms of actions on some horoballs (4.1.3). Section 4.2 proves the main density result (Theorem 4), following the ideas presented in [BM96, Section 8.3]. The main change with respect to [BM96] and [Moz99] is the replacement of local data by purely group-theoretic considerations (4.2.3). Section 4.3 recalls some facts about Moufang twin trees, and then describes the intersection of this theory with that of tree-lattices of Nagao type. This shows in particular that the class of non-uniform tree-lattices to which Theorem 5 applies is quite wide (4.3.4).

We thank Sh. Mozes for helpful discussions on Example 69 and are very grateful to the organizers of the conference ⟨⟨Geometric Group Theory⟩⟩ (Guwahati, Assam – India, December 2002) organized by the Indian Institute of Technology Guwahati and the Indian Statistical Institute (supported by the National Board of Higher Mathematics). The conference gave the authors the opportunity to complete their research on the subject which is presented in this paper.

4.1. Lattices of Nagao type

The tree lattices we are discussing in this paper are always fundamental groups of rays of groups. More precisely, we shall work with the following set-up.

Notation 9. Let R be an infinite ray with vertices $\{x_i\}_{i\geq 0}$. To each vertex x_i, $i \geq 0$, we attach a finite group Γ_i such that Γ_i is a subgroup of Γ_{i+1} for all $i > 0$. Furthermore, we are given a common subgroup H_0 of Γ_0 and Γ_1. We define the integers $k := [\Gamma_0 :$

$H_0]$, $q_0 := k - 1$, $q_1 := [\Gamma_1 : H_0]$ and $q_i := [\Gamma_i : \Gamma_{i-1}]$ for $i \geq 2$, and we require that $q_i \geq 2$ for all $i \geq 0$. Now we attach H_0 to the edge $\{x_0, x_1\}$ of R and Γ_i to $\{x_i, x_{i+1}\}$ for all $i > 0$. We have thus defined a graph of groups, where the graph is the ray R.

Picture 10.

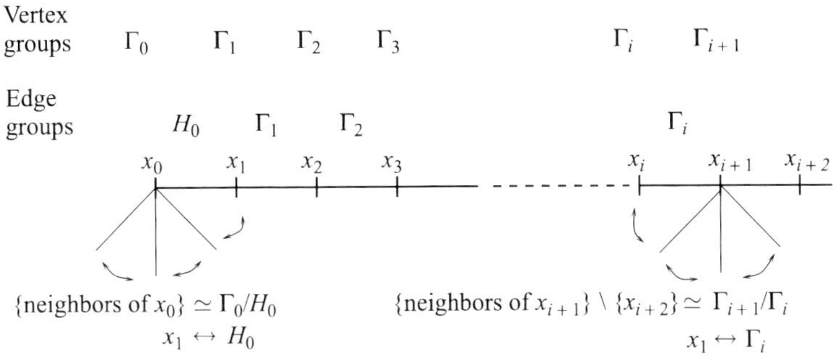

4.1.1. *Levels, degrees and a first density result*

Denote by Γ the fundamental group of this graph of groups and by T the canonical tree provided by Bass–Serre theory [Ser77, I.5.1]: Γ acts on T with R as a fundamental domain. For any $i \geq 0$, we identify Γ_i with the vertex stabilizer $\mathrm{Stab}_\Gamma(x_i)$ and $H_0 = \Gamma_0 \cap \Gamma_1$ with the edge stabilizer $\mathrm{Stab}_\Gamma(\{x_0, x_1\})$.

Definition 11. *Any vertex $x \in VT$ lies in a Γ-orbit $\Gamma \cdot x_i$ for precisely one $i \geq 0$; we call i the* level *of x and denote it by $\ell(x) = i$. We set $G := \mathrm{Aut}(T)$ and $L := \{g \in G | \ell(g \cdot x) = \ell(x)$ for all $x \in VT\}$.*

The kernel of the canonical homomorphism $\Gamma \to G$ is a (finite) subgroup of H_0, and by a slight abuse of notation, we shall denote the image of Γ in G again by Γ. Therefore Γ is a discrete subgroup of G, and it is of course also a subgroup of L. Our assumption $q_i \geq 2$ for all $i \geq 0$ implies that $\mathrm{Vol}(\Gamma \backslash\backslash T) = \sum_{i=0}^{\infty} |\Gamma_i|^{-1}$ is finite. Hence Γ is an (obviously non-uniform) T-lattice, called a *lattice of Nagao type* in [BL01, Chapter 10]. Since R is a fundamental domain for the action of Γ on T, we immediately obtain the following two facts.

Fact 12. Each vertex $x \in VT$ of level 0 has precisely $k = q_0 + 1$ neighbors, and all of them are of level 1. In particular, the degree (or valency) of x is $\deg(x) = k$. □

Fact 13. Each vertex $x \in VT$ with $\ell(x) = i > 0$ has precisely 1 neighbor of level $i + 1$ and q_i neighbors of level $i - 1$. In particular, $\deg(x) = q_i + 1$. □

Fact 13 immediately implies the following two statements.

Fact 14. If $\{y_j\}_{j \geq 0} \subset VT$ is a geodesic ray in T with $\ell(y_0) > \ell(y_1)$, then for any $j \leq \ell(y_0)$, we have: $\ell(y_j) = \ell(y_0) - j$. □

Fact 15. If $y \in VT$ has level $i > 0$, then there is a unique (infinite) ray $R(y) = (y_0, y_1, \ldots)$ starting in $y = y_0$ such that $\ell(y_j) = i + j$ for all $j \geq 0$. The ray $R(y)$ will be referred to as the level-increasing ray from x. □

It is well-known that either T is biregular (i.e. $\deg(x) = \deg(y)$ for any two vertices $x, y \in VT$ at even distance from each other) or $G\backslash T = R$ [BL01, Section 10.1]. However, we shall give a short elementary argument for this (in fact for a slightly more general) statement here, also because we shall later need the remark following this lemma.

Lemma 16. *If F is a subgroup of G which contains Γ and is not included in L, then $F \cdot x_0 = F \cdot x_{2m}$ and $F \cdot x_1 = F \cdot x_{2m+1}$ for all integers $m \geq 0$. Hence in this case T has to be biregular, and F acts transitively on vertices of the same type (i.e. at even distance from each other) and on the set ET of (geometric) edges of T.*

Proof. Since for any $x \in VT$, $\ell(x)$ is equal to the distance of x from the orbit $\Gamma \cdot x_0$, F acts level-preservingly on T if it stabilizes $\Gamma \cdot x_0$. Since F is not included in L, there is an $n > 0$ with $F \cdot x_0 = F \cdot x_n$. We choose $n > 0$ minimal with this property. If $n = 1$, we inductively get $F \cdot x_r = F \cdot x_{r-1} = F \cdot x_0$ for all $r \in \mathbf{N}$. If $n = 2$, then inductively $F \cdot x_3 = F \cdot x_1$, $F \cdot x_4 = F \cdot x_0 (= F \cdot x_2)$, $F \cdot x_5 = F \cdot x_1$ and so on.

Now assume that $n > 2$. Since $F \cdot x_{n-1} = F \cdot x_1$, we can choose a $g \in F$ with $g \cdot x_{n-1} = x_1$. Since $g \cdot x_n \neq x_2$ (otherwise $F \cdot x_2 = F \cdot x_n = F \cdot x_0$, contradicting the minimality of n), $\ell(g \cdot x_n) = 0$ by Fact 13. Now x_1 has $q_1 \geq 2$ neighbors of level 0, and Γ_1 acts transitively on them. Hence we can choose $\gamma_1 \in \Gamma_1$ such that $\gamma_1 \cdot (g \cdot x_n) \neq g \cdot x_n$. Hence $f := g^{-1}\gamma_1 g$ is in $F_{x_{n-1}}$ and $f \cdot x_n \neq x_n$. Therefore $\Gamma_{n-1} f \cdot x_n$ contains x_{n-2}, and thus $F \cdot x_{n-2} = F \cdot x_n = F \cdot x_0$. However, this again contradicts the minimality of n, and so $n > 2$ is impossible.

This proves the first claim, which immediately shows that F acts transitively on vertices of the same type. In particular, for any edge $e \in ET$, there is an $f \in F$ such that $x_0 \in f \cdot e$, and then there is a $\gamma_0 \in \Gamma_0$ such that $(\gamma_0 \cdot f) \cdot e = \{x_0, x_1\}$. Hence F acts transitively on ET. □

Remark 17. It is not needed in the above argument that R is infinite. So if we have the same set-up as above, only with a finite path R' instead of R, and if T' and G' are defined similarly, then also T' has to be biregular or else $G'\backslash T' = R'$.

Lemma 18. *Any level-preserving isomorphism $\psi : T_1 \to T_2$ between two subtrees T_1 and T_2 of T extends to an element of L.*

Proof. By Zorn's lemma it suffices to show that ψ can always be properly and level-preservingly extended if $T_1 \neq T$. That this is in fact true can readily be checked by using the Facts 12 and 13 above. □

Remark 19. This lemma in particular shows that the group L is non-discrete and uncountable.

The elementary statements which we deduced above have an interesting consequence if we combine them with an idea presented in the last paragraph of Mozes' paper [Moz99]. Recall that G° denotes the group of type-preserving automorphisms of T.

Proposition 20. *If F is a subgroup of G properly containing L (so that T must be biregular), then the closure of F in G contains G°.*

Proof. We have to show that any type-preserving isomorphism $\phi : S_1 \to S_2$ between two finite subtrees S_1 and S_2 of T can be extended to an element of F. Without loss of generality, we may assume that $S_1 = B_s(z_1)$, $S_2 = B_s(z_2)$ are two balls of radius $s \in \mathbb{N}$ with centers z_1 and z_2 which are of the same type. Let v_1 be a terminal vertex in S_1 and set $v_2 = \phi(v_1)$. So also v_1 and v_2 are of the same type. Let u_j $(j = 1, 2)$ be the unique neighbor of v_j in S_j. Choose a vertex $x_n \in R$ of the same type as v_1 and v_2 with $n \geq 2s$. By Lemma 16, there exist elements $f_j \in F$ satisfying $f_j \cdot v_j = x_n$ and $f_j \cdot u_j = x_{n-1}$ for $j = 1, 2$. Set $T_j := f_j(S_j)$ for $j = 1, 2$, and consider the isomorphism $\psi := f_2|_{S_2}\phi f_1{}^{-1}|_{T_1} : T_1 \to T_2$. Note that ψ fixes x_n and x_{n-1} by construction. So Fact 14 together with $n \geq 2s$ implies that ψ is automatically level-preserving. Hence by Lemma 18, there exists a $g \in L$ satisfying $g|_{T_1} = \psi$. Therefore the element $f := f_2{}^{-1} g f_1 \in F$ extends ϕ. □

4.1.2. *Lattices of directly split Nagao type*

Recall that we want to investigate the commensurator $C(\Gamma) = \{g \in G | \Gamma \cap g\Gamma g^{-1}$ has finite index in Γ and in $g\Gamma g^{-1}\}$. In the present generality, we even cannot exclude the possibility that $C(\Gamma) = \Gamma$ (for such an example, see [BL01, Section 10.5]). Therefore we have to consider a restricted class of lattices of Nagao type. Motivated by and generalizing the discussion in [BM96, Section 8.3], we choose the following one.

Definition 21. *We say that the above introduced tree lattice* Γ *is of* directly split Nagao type *if there exist, for all* $j > 0$, H_0*-invariant subgroups* $U_j \leq \Gamma_j$, *called* root groups, *such that* $\Gamma_i = H_0 \ltimes (U_1 \times U_2 \times \cdots \times U_i)$ *for all* $i > 0$.

Remark 22.

1. ⟨⟨Directly split⟩⟩ above refers to the direct products $U_1 \times U_2 \times \cdots \times U_i$. In view of this terminology one would perhaps expect that also the product with H_0 is direct. However, this would be too restrictive; it would even exclude the classical Nagao lattice $\mathrm{PSL}_2(\mathbf{F}_q[t^{-1}])$.
2. The terminology ⟨⟨root group⟩⟩ is motivated by building theory (here: twin tree theory) and will be explained in Section 4.3.
3. If $H_0 = \{1\}$, our lattices of directly split Nagao type are the same as the *product groupings* associated to Nagao rays in [BL01, Section 10.6].
4. We stress again that our root groups U_i can be completely arbitrary non-trivial finite groups.
5. It would be desirable, especially in view of Moufang twin trees which do not satisfy the Condition (Comm), to obtain results as deduced below for splittings $\Gamma_i = H_0 \ltimes (U_1 \ldots U_i)$ which are not necessarily direct. However, our method really requires that the different root groups commute. We shall give a geometric characterization of this algebraic property in Lemma 31 below.

For the rest of this paper, we shall assume that our lattices are of directly split Nagao type. We start by collecting a few elementary statements, the first two of which are obvious.

Fact 23. We have: $\Gamma_1 = H_0 \ltimes U_1$ and $\Gamma_i = \Gamma_{i-1} \ltimes U_i$ for any $i \geq 2$. In particular, $|U_i| = q_i \geq 2$ for any $i \geq 1$. □

Fact 24. For any $i > 0$, the group U_i acts simply transitively (i.e. regularly) on the set of neighbors of x_i which are different from x_{i+1}. □

This can easily be generalized as follows.

Lemma 25. *For any two integers i and j with* $0 < i \leq j$, *the group* $U_{i,j} := U_i \times U_{i+1} \cdots \times U_j$ *acts simply transitively on* $M_{i,j} := \{x \in VT | \ell(x) = i - 1$ *and* $\mathrm{dist}(x, x_j) = j - i + 1\}$, *where* $\mathrm{dist}(x, x_j)$ *denotes the distance between* x *and* x_j *in* T.

Proof. The fact that $U_{i,j}$ acts transitively on $M_{i,j}$ immediately follows from Fact 24 by induction on $j - i$. Now assume that an element $u = u_j \ldots u_i \in U_{i,j}$ fixes x_{i-1}. Since $U_{i,j} \leq \Gamma_j$, u has to fix x_l for all $i - 1 \leq l \leq j$. In particular, $x_{j-1} = u \cdot x_{j-1} = u_j \cdot x_{j-1}$, which implies $u_j = 1$, again by Fact 24. Going on this way, we get $u_j = u_{j-1} = \cdots = u_i = 1$. □

Picture 26.

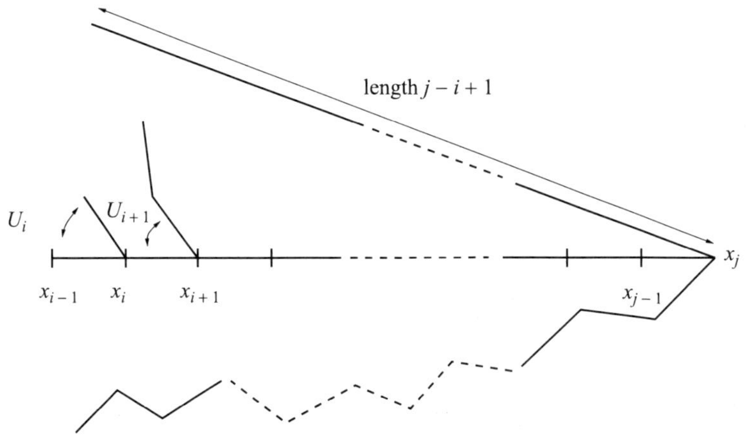

4.1.3. *Horoballs and horospheres*

Recall that to any end ϵ of the tree T, we can associate horospheres and horoballs centered at ϵ, see for instance [BL01, Section 9.2]. In the present paper, only certain horospheres and horoballs, defined by vertices of positive level, will play a role.

Definition 27. *Let x be a vertex of level* $\ell(x) > 0$, *and let* ϵ_x *be the end of T defined by the level-increasing ray from x (see Fact 15).*

(a) *We denote by* $\mathrm{HS}(x)$ *the horosphere centered at* ϵ_x *which contains* x.
(b) *We denote by* $\mathrm{HB}(x)$ *the horoball centered at* ϵ_x *which has* $\mathrm{HS}(x)$ *as its boundary, i.e. as its set of terminal vertices.*

Remark 28. It is easy to see that the horoball $\mathrm{HB}(x)$ is the connected component of x in the forest spanned by all vertices of level $\geq \ell(x)$, and that the horosphere $\mathrm{HS}(x)$ is the set of vertices of level $\ell(x)$ in $\mathrm{HB}(x)$.

The following statement is an immediate consequence of Lemma 25 and the above definition.

Fact 29. For any integer $i > 0$, the subgroup $W_{i+1} := \langle U_j | j \geq i+1 \rangle$ of Γ acts simply transitively on the horosphere $\mathrm{HS}(x_i)$. □

Remark 30. As opposed to Lemma 31 below, this fact remains true in a more general context (for instance for general Moufang twin trees), where the products $U_1 \dots U_i$ are not required to be direct.

We close this section by stating a geometric consequence of the directness of these products which will play an important role in the next section.

Lemma 31. *For any $i > 0$, the group U_i fixes the horoball* $\mathrm{HB}(x_i)$ *pointwise.*

Proof. It suffices to show that U_i acts trivially on $\mathrm{HS}(x_i)$. So take an arbitrary vertex $x \in \mathrm{HS}(x_i)$, and let w be the unique element in W_{i+1} satisfying $x = w \cdot x_i$ (Fact 29). Now since U_i commutes with U_j for all $j > i$, it also commutes with W_{i+1}. Therefore $u_i \cdot x = u_i \cdot (w \cdot x_i) = w \cdot (u_i \cdot x_i) = w \cdot x_i = x$ for all $u_i \in U_i$. □

4.2. Proof of the main density theorem

We are now going to investigate the commensurator $C(\Gamma)$, where Γ is a T-lattice of directly split Nagao type as introduced in Section 4.1. Let us first of all mention that Question 1 formulated in the introduction has a positive answer in this case. In fact, according to [BL01, Section 10.1], Γ is already of infinite index in $N_G(\Gamma)$, its normalizer in G. In this section, it is our goal to show that $C(\Gamma) \cap L$ is dense in L, which is the major step towards answering Question 2. Our strategy will be an appropriate adaptation (using more group theoretic constructions instead of ⟨⟨local data⟩⟩) of the method developed in [BM96, Section 8.3]. In particular, we shall also prove an extension result for commensurators of certain uniform tree lattices for subtrees of T with bounded level function (see Proposition 54) and then apply Liu's theorem ([Liu94]).

As in [Moz99], it will be technically convenient to work with a certain finite index subgroup Δ of Γ (so that of course $C(\Delta) = C(\Gamma)$) instead of Γ itself. In the case of the classical Nagao lattice $\mathrm{PSL}_2(\mathbf{F}_q[t^{-1}])$, Sh. Mozes used the first congruence subgroup. In the theory of Kac-Moody groups, one would analogously choose Δ to be the ⟨⟨unipotent radical⟩⟩ of Γ, where Γ is considered as a parabolic subgroup of a Kac-Moody group over a finite field. In the next subsection we shall define Δ in terms of the subgroups U_i introduced in Definition 21.

4.2.1. *The group Δ and its action on the tree and on horoballs*

We keep the notations introduced in Section 4.1 and define further subgroups of the T-lattice Γ of directly split Nagao type. First, we call $V := \langle U_i | i > 0 \rangle = \bigoplus_{i>0} U_i$ the direct sum of the groups U_i when i ranges over all the positive integers i. We fix a system $\{\gamma_1 := 1; \gamma_2; \dots; \gamma_k\}$ of representatives of the cosets in Γ_0/H_0. For $i > 0$ and $1 \le s \le k$, we also set:

$$U_{i,s} := \gamma_s U_i \gamma_s^{-1} \quad \text{and} \quad V_s := \langle U_j, s | j > 0 \rangle = \gamma_s V \gamma_s^{-1},$$

so that $V_1 = V$, and we finally define $\Delta := \langle V_s | 1 \le s \le k \rangle$.

Lemma 32. *The group Δ is the normal closure of V in Γ, and we have: $\Gamma = \Gamma_0 \ltimes \Delta$.*

Proof. Recall that $\Gamma = \langle \Gamma_0 \cup V \rangle$ and that H_0 normalizes V. This implies $\Gamma = \Gamma_0 \Delta$ (since $V\Gamma_0\Delta = \Gamma_0\Delta$), and hence Δ is normal in Γ.

The definition of Γ as the fundamental group of a ray of groups implies that Γ is an amalgam of the form $\Gamma = \Gamma_0 *_{H_0} H_0 V$. Now the normal form for amalgams (as stated for instance in [Ser77, I.1.2]) shows that no product of the form $\gamma_{j_1} v_1 \gamma_{j_1}^{-1} \gamma_{j_2} v_2 \gamma_{j_2}^{-1} \dots \gamma_{j_l} v_l \gamma_{j_l}^{-1}$ with $l \in \mathbb{N}$, $v_1, \dots, v_l \in V \backslash \{1\}$, $1 \le j_r \le k$ for all $r \le l$ and $j_r \ne j_{r+1}$ (hence $\gamma_{j_r}^{-1} \gamma_{j_{r+1}} \notin H_0$) for all $r < k$ is an element of Γ_0. Hence $\Gamma_0 \cap \Delta = \{1\}$. □

The same normal form argument also shows that $\Delta = V_1 * V_2 * \dots * V_k$.

Lemma 33. *For any positive integer $i > 0$, the stabilizer $\mathrm{Stab}_\Delta(x_i)$ is equal to $U_1 \times U_2 \times \dots \times U_i$ and acts trivially on $\mathrm{HB}(x_i)$.*

Proof. Since $H_0 \cap \Delta = \{1\}$ by the previous lemma, we obtain $\mathrm{Stab}_\Delta(x_i) = \Gamma_i \cap \Delta = (H_0 \ltimes (U_1 \times U_2 \times \dots \times U_i)) \cap \Delta = U_1 \times U_2 \times \dots \times U_i$. The second assertion now follows from Lemma 31. □

Remark 34. The triviality of the $\mathrm{Stab}_\Delta(x_i)$-action on $\mathrm{HB}(x_i)$ is the main reason why we work with Δ instead of Γ: we do not have the same property for Γ_i since H_0 usually acts non-trivially on $\mathrm{HB}(x_i)$.

Lemma 35. *The subtree $F := \Gamma_0 \cdot R$ is a fundamental domain for the Δ-action on T.*

Proof. We have: $\Delta \cdot F = (\Delta \cdot \Gamma_0) \cdot R = \Gamma \cdot R = T$. Now assume that there are $\delta \in \Delta$ and $g, h \in \Gamma_0$ and $i, j \ge 0$ such that $\delta \cdot (g \cdot x_i) = h \cdot x_j =: x$. Then first of all $i = j = \ell(x)$. Secondly, we may and shall assume $i > 0$ (otherwise $\delta = 1$ by Lemma 32). So $h^{-1}\delta g$ is in $\Gamma_i \subseteq \Delta H_0$. Since $h^{-1}\delta h \in \Delta$, we conclude that $h^{-1}g$ is also in ΔH_0. However, $\Gamma_0 \cap \Delta \cdot H_0 = H_0$ by Lemma 32, so that $h^{-1}g \in H_0$. Therefore: $g \cdot x_i = h \cdot x_i = h \cdot x_j$. □

This lemma shows in particular that Δ acts simply transitively on the set of all vertices of level 0. Using again the coset representatives $\{\gamma_1 := 1; \gamma_2; \dots; \gamma_k\}$ for Γ_0/H_0, we see that F is the union of the k rays $R_s := \gamma_s \cdot R$ with $1 \le s \le k$. We number the vertices of R_s by $x_{i,s} := \gamma_s \cdot x_i$ for all $i \ge 0$ (so $x_{0,s} = x_0$ for all s and $x_{i,1} = x_i$ for all i).

Claim 36. For any $1 \le s \le k$ and $i > 0$, we have:

(i) The group $\langle U_{j,s} | j > i \rangle$ acts simply transitively on $\mathrm{HS}(x_{i,s})$.
(ii) The stabilizer $\mathrm{Stab}_\Delta(x_{i,s})$ is equal to $U_{1,s} \times \dots \times U_{i,s}$ and acts trivially on $\mathrm{HB}(x_{i,s})$.

Proof. The first statement follows directly from Fact 29, and the second from $\mathrm{Stab}_\Delta(x_{i,s}) = \gamma_s \mathrm{Stab}_\Delta(x_i)\gamma_s^{-1}$ (here we use that Δ is normal in Γ) and Lemma 33. □

Lemma 37. *For any vertex x of level > 0, we have:*

(i) *the group* $\mathrm{Stab}_\Delta(x)$ *acts trivially on* $\mathrm{HB}(x)$*;*
(ii) *if δ, δ' are two elements in Δ with $\delta \cdot x = \delta' \cdot x$, then $\delta|_{\mathrm{HB}(x)} = \delta'|_{\mathrm{HB}(x)}$.*

Proof. (ii) is a straightforward consequence of (i), and (i) follows from combining Claim 36 and Lemma 35. □

Given $x \in VT$ with $\ell(x) > 0$, we can now label the elements in $\mathrm{HS}(x)$ by elements of Δ.

Definition 38.

(a) *For $x = x_{i,s} \in F$ with $1 \le s \le k$ and $i > 0$ and for any $y \in \mathrm{HS}(x)$, we define $\delta_{x,y}$ to be the unique element in $\langle U_{j,s} | j > i\rangle$ satisfying $\delta_{x,y} \cdot x = y$ (see Claim 36).*
(b) *For arbitrary $x \in VT$ with $\ell(x) > 0$ and $y \in \mathrm{HS}(x)$, we choose $\delta \in \Delta$ such that $\delta \cdot x \in F$ and set $\delta_{x,y} := \delta^{-1}\delta_{\delta(x),\delta(y)}\delta$.*

The definition in part (b) above makes sense thanks to:

Claim 39. The above introduced element $\delta_{x,y}$ is well-defined.

Proof. Assume that also $\delta' \cdot x \in F$ for some $\delta' \in \Delta$ and set: $z := \delta' \cdot x = x_{i,s} = \delta \cdot x$ (see Lemma 35). Then $\delta'' := \delta'\delta^{-1} \in \mathrm{Stab}_\Delta(z) = U_{1,s} \times \cdots \times U_{i,s}$. So firstly, δ'' acts trivially on $\mathrm{HS}(z) = \mathrm{HS}(\delta \cdot x) = \delta \cdot \mathrm{HS}(x)$ by Lemma 37, hence $\delta' \cdot y = \delta'' \cdot (\delta \cdot y) = \delta \cdot y$; and secondly δ'' commutes with $\langle U_{j,s} | j > i\rangle$, in particular with $\delta_{z,\delta(y)}$. (Here we use again that Γ is of *directly* split type.) Therefore $\delta'^{-1}\delta_{\delta'(x),\delta'(y)}\delta' = \delta^{-1}\delta''^{-1}\delta_{z,\delta(y)}\delta''\delta = \delta^{-1}\delta_{\delta(x),\delta(y)}\delta$. □

Having established this, it is easy to check the following calculation rules:

Lemma 40. *For $x \in VT$ with $\ell(x) > 0$ and $y \in \mathrm{HS}(x)$, we have:*

(i) *$\delta_{x,y}$ is in Δ and $\delta_{x,y} \cdot x = y$;*
(ii) *$h\delta_{x,y}h^{-1} = \delta_{h(x),h(y)}$ for all $h \in \Delta$;*
(iii) *$\delta_{y,x} = \delta_{x,y}^{-1}$;*
(iv) *$\delta_{y,z}\delta_{x,y} = \delta_{x,z}$ for all $z \in \mathrm{HS}(x)$.*

Proof. (i) is clear by definition. To prove (ii), we note that if $\delta \in \Delta$ is such that $\delta \cdot x \in F$, then we can choose $\delta h^{-1} \in \Delta$ in the definition of $\delta_{h(x),h(y)}$ in order to have $(\delta h^{-1}) \cdot (h \cdot x) = \delta \cdot x \in F$. Therefore, applying part (b) of Definition 38, we obtain: $\delta_{h(x),h(y)} = (\delta h^{-1})^{-1}\delta_{\delta(x),\delta(y)}(\delta h^{-1}) = h\delta_{x,y}h^{-1}$.

According to (ii), it is enough to prove (iii) and (iv) for $x \in F$. So assume $x = x_{i,s}$ with $i > 0$ and $1 \le s \le k$. Set $\delta := \delta_{x,y}^{-1}$, and observe that this is an element of $\langle U_{j,s} | j > i\rangle$ sending y to $x \in F$ as well as $\delta \cdot y = x$ to $\delta \cdot x$. So by part (a) of

Definition 38, $\delta = \delta_{\delta(y),\delta(x)}$. Therefore, now by part (b) of Definition 38, $\delta_{y,x} = \delta^{-1}\delta_{\delta(y),\delta(x)}\delta = \delta^{-1}\delta\delta = \delta = \delta_{x,y}^{-1}$, proving (iii). In order to verify (iv), we first note that, again by part (b) of Definition 38, $\delta_{y,z} = \delta^{-1}\delta_{\delta(y),\delta(z)}\delta = \delta^{-1}\delta_{x,\delta(z)}\delta$, which is an element of $\langle U_{j,s} | j > i \rangle$. Hence also $\delta_{y,z}\delta_{x,y}$ is an element of $\langle U_{j,s} | j > i \rangle$, and by (i), it sends x to z. Therefore, by part (a) of Definition 38, $\delta_{y,z}\delta_{x,y} = \delta_{x,z}$, proving (iv). □

4.2.2. *Connected components of bounded level*

We are now going to generalize the strategy of [BM96, Section 8.3] to the present situation. We recall the definition of a certain graph G_i associated to T and a given level $i > 0$.

Definition 41. *For $i > 0$, we denote by T_i the subforest of T spanned by all vertices of level $\leq i$. For any vertex x with $\ell(x) \leq i$, we denote by $C_i(x)$ the connected component of T_i containing x. We set*

$$\mathcal{C}_i := \{C_i(x) | x \in VT \text{ and } \ell(x) \leq i\},$$

and define a (simplicial) graph

$$\mathrm{G}_i = (V\mathrm{G}_i, E\mathrm{G}_i) \quad \text{with vertex set} \quad V\mathrm{G}_i = \mathcal{C}_i$$

by declaring a two element subset $\{X, Y\}$ of $V\mathrm{G}_i$ to be in the edge set $E\mathrm{G}_i$ if and only if there exist vertices $x \in X$, $y \in Y$ such that $\ell(x) = \ell(y) = i$ and $y \in \mathrm{HS}(x)$.

Remark 42.

1. If $\{X, Y\} \in E\mathrm{G}_i$, the vertices $x \in X$ and $y \in Y$ in the above definition are uniquely determined: they are the unique terminal vertices of X and Y, respectively, contained in any geodesic in T which starts in a vertex of X and ends in a vertex of Y.
2. If $X, Y \in \mathcal{C}_i$ are given, there is a unique geodesic in G_i connecting them. It is obtained from an arbitrary geodesic p in T which connects a vertex of X to a vertex of Y by first intersecting p with T_i, yielding a disjoint union p' of geodesics in T_i, then replacing these subgeodesics of p' with their corresponding connected components in $\mathcal{C}_i$ (considered as elements of $V\mathrm{G}_i$) and connecting the adjacent ones among them by edges in $E\mathrm{G}_i$.
3. A path $\mathcal{P}$ in G_i, i.e. a finite sequence $\mathcal{P} = (X_0, \ldots, X_n)$ with $X_j \in \mathcal{C}_i$ for all $j \leq n$ and $\{X_j, X_{j+1}\} \in E\mathrm{G}_i$ for all $j < n$, is a geodesic in G_i if and only if X_{j+2} is neither equal nor adjacent to X_j for all $j \leq n-2$. In fact, there is only one path $\mathcal{P}$ between X_0 and X_n in G_i with this last mentioned property. This follows from the uniqueness of geodesics in T.

The elements $\delta_{x,y}$ introduced in the previous subsection now provide us with transformations $\tau_{X,Y}$ mapping $X \in \mathcal{C}_i$ onto $Y \in \mathcal{C}_i$.

Definition 43. *Let $X, Y \in \mathcal{C}_i$ be given.*

(a) *If X and Y are adjacent in G_i, and if $x \in VX$, $y \in VY$ are such that $y \in \mathrm{HS}(x)$, we set $\tau_{X,Y} := \delta_{x,y}$. This is well-defined by Remark 42 (1).*

(b) *If X and Y are arbitrary, we consider the unique geodesic $(X = X_0, X_1, \ldots, X_n = Y)$ from X to Y in G_i, and set $\tau_{X,Y} := \tau_{X_{n-1},X_n}\tau_{X_{n-2},X_{n-1}} \cdots \tau_{X_0,X_1}$. We also set $\tau_{X,X} := 1$.*

Remark 44. We note that $\tau_{X,Y} = \tau_{Z_{m-1},Z_m}\tau_{Z_{m-2},Z_{m-1}}, \ldots, \tau_{Z_0,Z_1}$ is true for *any* path $\mathcal{P} = (X = Z_0, Z_1, \ldots, Z_m = Y)$ connecting X and Y in G_i. Indeed, if $\mathcal{P}$ is not a geodesic, then by Remark 42 (3), there exists a subpath (A, B, C) of $\mathcal{P}$ such that $A = C$ or $\{A, C\} \in E\mathrm{G}_i$. In the first case, we delete B and $C = A$ from $\mathcal{P}$ and observe that $\tau_{B,A}\tau_{A,B} = 1$ by Lemma 40 (iii). In the second case, we delete B from $\mathcal{P}$ and note that $\tau_{B,C}\tau_{A,B} = \tau_{A,C}$ by Lemma 40 (iv). So simultaneously shortening the path $\mathcal{P}$ and the product of τ's associated to it without changing the value of this product, we finally obtain the geodesic between X and Y in G_i, and the associated product is precisely the one occurring in the above definition of $\tau_{X,Y}$.

Lemma 45. *For any $X, Y, Z \in \mathcal{C}_i$, we have:*

(i) $\tau_{X,Y} \in \Delta$ *and* $\tau_{X,Y}(X) = Y$;
(ii) $h\tau_{X,Y}h^{-1} = \tau_{h(X),h(Y)}$ *for all* $h \in \Delta$;
(iii) $\tau_{Y,X} = \tau_{X,Y}^{-1}$;
(iv) $\tau_{Y,Z}\tau_{X,Y} = \tau_{X,Z}$.

Proof. The properties (i) to (iii) follow directly from the corresponding statements in Lemma 40, and (iv) is a consequence of Remark 44. □

We close this subsection by introducing truncated versions of T, R, F and Δ. For $i > 0$, we set $Y_i := C_i(x_0)$, $R^{(i)} := R \cap Y_i$, $F^{(i)} := F \cap Y_i = \Gamma_0 \cdot R^{(i)}$ and

$$\Delta_i := \langle U_{j,s} | 1 \leq j \leq i \quad \text{and} \quad 1 \leq s \leq k \rangle.$$

Lemma 46.

(i) *$R^{(i)}$ is a fundamental domain for the action of $\langle \Gamma_0 \cup \Gamma_i \rangle = \mathrm{Stab}_\Gamma(Y_i)$ on Y_i;*
(ii) *$F^{(i)}$ is a fundamental domain for the action of $\Delta_i = \mathrm{Stab}_\Delta(Y_i)$ on Y_i;*
(iii) *Δ_i (or more precisely: its image in $\mathrm{Aut}(Y_i)$) is a uniform Y_i-lattice.*

Proof. Set $\Gamma' := \langle \Gamma_0 \cup \Gamma_i \rangle$, which is obviously contained in $\mathrm{Stab}_\Gamma(Y_i)$. It is clear that $\Gamma' \cdot R^{(i)}$ contains a neighborhood of $R^{(i)}$ in Y_i. Thus $\Gamma' \cdot R^{(i)}$ is open in Y_i. But as a subgraph of Y_i, it is also closed. Since Y_i is connected, we conclude $\Gamma' \cdot R^{(i)} = Y_i$. (Side remark: this kind of reasoning already occurs in [Ser77].) Thus $\Gamma' \backslash Y_i = R^{(i)}$. If $\gamma \in \Gamma$ stabilizes Y_i, then $\gamma \cdot x_j \in \Gamma' \cdot x_j$ for any $j \leq i$, and since $\mathrm{Stab}_\Gamma(x_j) = \Gamma_j \subseteq \Gamma'$, we obtain $\gamma \in \Gamma'$. This proves (i).

The proof of (ii) is similar. We just observe that also $\Delta_i \cdot F^{(i)}$ (which contains a neighborhood of x_0 by definition of $F^{(i)}$) is open in Y_i due to the transitivity properties of the $U_{j,s}$ (see Fact 24). This implies $\Delta_i \cdot F^{(i)} = Y_i$ and hence $\Delta_i \backslash Y_i = F^{(i)}$. Since $\mathrm{Stab}_\Delta(x) \subseteq \Delta_i$ for all vertices $x \in F^{(i)}$ (Claim 36 (ii)), we also obtain $\Delta_i = \mathrm{Stab}_\Delta(Y_i)$. Finally, (iii) is an immediate consequence of (ii). □

4.2.3. *Permuting elements of the same level*

We now replace the analysis of ⟨⟨local data⟩⟩ which were used in [BM96] and [Moz99] by a more group theoretic construction. The latter is based on the observation Lemma 37 (ii) above and similar to the definition of the $\delta_{x,y}$. However, one has to be a bit careful here since Δ does not act transitively on the set of vertices of a given level.

Definition 47. *For any given vertex x of level $i > 0$, let $s = s(x)$ be the unique integer with $1 \leq s \leq k$ and $x \in \Delta \cdot R_s$ (see Lemma 35).*

(a) *We choose once and for all a $\delta_x \in \Delta$ such that $x = \delta_x \cdot x_{i,s} = (\delta_x \gamma_s) \cdot x_i$, and we set: $\gamma_x := \delta_x \gamma_s$.*
(b) *For any two vertices x, y of the same level $i > 0$, we define: $\gamma_{x,y} := \gamma_y \gamma_x^{-1}$.*

Note that we work with the fixed system of coset representatives $\{\gamma_1 := 1; \gamma_2; \ldots; \gamma_k\}$ for Γ_0/H_0 and that $\gamma_{x,y}$ need not coincide with $\delta_{x,y}$ in case x and y are in the same horosphere (see, however, (iv) below which we will need to prove Claim 52).

Lemma 48. *For vertices x, y, z of the same level $i > 0$, we have:*

(i) *$\gamma_{x,y}$ is in Γ and $\gamma_{x,y} \cdot x = y$;*
(ii) *$\gamma_{y,x} = \gamma_{x,y}^{-1}$;*
(iii) *$\gamma_{y,z}\gamma_{x,y} = \gamma_{x,z}$;*
(iv) *if $y \in \Delta \cdot x$ then $\gamma_{x,y} \in \Delta$.*

Proof. (i), (ii) and (iii) are obvious by definition. If $x \in \Delta \cdot R_s$ and $y \in \Delta \cdot R_t$ with $1 \leq s, t \leq k$, then $\gamma_{x,y} \in \Delta\gamma_t\gamma_s^{-1}\Delta = \Delta\gamma_t\gamma_s^{-1}$ because $\Delta \lhd \Gamma$ (Lemma 32). This implies (iv). □

The next property is a bit more subtle. It will be needed in order to obtain a reasonable definition of the groups L_i (see Definition 50 and Remark 51 (1)) which are essential for the extensions to be constructed in the next subsection.

Lemma 49. *If x, y are vertices of level $i > 0$, $x' \in \mathrm{HS}(x)$ and $y' := \gamma_{x,y}(x')$, then we have: $\gamma_{x,y}|_{\mathrm{HB}(x)} = \gamma_{x',y'}|_{\mathrm{HB}(x)}$.*

Proof. By Lemma 48 (iii), we have: $\gamma_{x',y'} = \gamma_{y,y'}\gamma_{x,y}\gamma_{x',x}$. Since $x' \in \mathrm{HS}(x)$ and hence $y' \in \gamma_{x,y} \cdot \mathrm{HS}(x) = \mathrm{HS}(y)$, $x' \in \Delta \cdot x$ and $y' \in \Delta \cdot y$ by the transitive action of Δ on $\mathrm{HS}(x)$, respectively $\mathrm{HS}(y)$ (combine Claim 36 and Lemma 35). Thus Lemma 48 (iv) tells us that $\gamma_{x',x}$ and $\gamma_{y,y'}$ are in Δ. Therefore $\gamma_{x',y'} \in \Delta\gamma_{x,y}\Delta = \gamma_{x,y}\Delta$, the last equality again following from $\Delta \lhd \Gamma$. Hence there exists a $\delta \in \Delta$ such that $\gamma_{x',y'} = \gamma_{x,y}\delta$. Since $\delta \cdot x' = \gamma_{x,y}^{-1}\gamma_{x',y'} \cdot x' = \gamma_{x,y}^{-1}(y') = x'$, we have $\delta \in \mathrm{Stab}_\Delta(x')$. Therefore δ acts trivially on $\mathrm{HB}(x') = \mathrm{HB}(x)$ (Lemma 37), and finally: $\gamma_{x',y'}|_{\mathrm{HB}(x)} = \gamma_{x,y}\delta|_{\mathrm{HB}(x)} = \gamma_{x,y}|_{\mathrm{HB}(x)}$. □

4.2.4. *Extensions and commensurators*

We can now define suitable extensions of isomorphisms between subtrees $X, Y \in \mathcal{C}_i$ of T (see Proposition 53 below). For this we need to define certain subgroups of L.

Definition 50. *Let $i > 0$ be a positive integer. By L_i we denote the group of all level-preserving automorphisms $h \in \mathrm{Aut}(T)$ satisfying:*

(a) *for any vertex x with $\ell(x) = i$, $h|_{\mathrm{HB}(x)} = \gamma_{x,h(x)}|_{\mathrm{HB}(x)}$;*
(b) *for all $X, Y \in \mathcal{C}_i$, $(h\tau_{X,Y}h^{-1})|_{h(X)} = \tau_{h(X),h(Y)}|_{h(X)}$.*

Remarks 51.

1. L_i is indeed a subgroup of L. If $g, h \in L_i$ then gh, h^{-1} clearly satisfy (b), and it follows from Lemma 48 that they also satisfy (a). In the extension procedure, we only need property (a) of the above definition at certain vertices x. However, one would not get a reasonable definition of L_i if one did not require (a) for all level i vertices x. But then one has to check that the different vertices in $\mathrm{HS}(x)$ do not lead to different conditions for $h|_{\mathrm{HB}(x)}$: this is precisely what we established in Lemma 49.
2. It is sufficient to require Condition (b) for a *fixed* $X_0 \in \mathcal{C}_i$ (and all $Y \in \mathcal{C}_i$). This follows from Lemma 45: $\tau_{X,Y} = \tau_{X_0,Y}\tau_{X_0,X}^{-1}$ for all $X, Y \in \mathcal{C}_i$.

Claim 52. For all $i > 0$, the group Δ is contained in L_i.

Proof. Condition (a) for elements of Δ follows from combining Lemma 37 (ii) and Lemma 48 (iv). Condition (b) follows from Lemma 45 (ii). □

Up to replacing local data by the elements $\gamma_{x,y}$, the next proposition is similar to [BM96, Proposition 8.5].

Proposition 53. *For any two $X, Y \in \mathcal{C}_i$ and any level-preserving isomorphism $h : X \to Y$, there is a unique extension of h to an automorphism $E(h) \in L_i$.*

Proof. We first show how we can (and must) extend h to horoballs and elements of $\mathcal{C}_i$ which are neighboring X.

(1) *Definition of $E(h)$ on* $\mathrm{HB}(x)$. According to Condition (a) of Definition 50, we have only one choice to define $E(h)$ on $\mathrm{HB}(x)$ for $x \in VX$ with $\ell(x) = i$, namely by $E(h)|_{\mathrm{HB}(x)} = \gamma_{x,y}|_{\mathrm{HB}(x)}$, where $y := h(x)$.
(2) *Definition of $E(h)$ on Z.* For $x \neq z \in \mathrm{HS}(x)$, we set $Z := C_i(z) \in \mathcal{C}_i$, $z' := E(h)(z) = \gamma_{x,y}(z)$ and $Z' := C_i(z')$. Note that $z' \in \mathrm{HS}(y)$. According to Condition (b) of Definition 50, we have to define $E(h)$ on Z by $E(h)|_Z = \tau_{Y,Z'}h\tau_{Z,X}|_Z$.
(3) *Equality of the two definitions on* $\{z\} = Z \cap \mathrm{HB}(x)$. Since $z' \in \mathrm{HS}(y)$ and $z \in \mathrm{HS}(x)$, Definition 43 implies $\tau_{Y,Z'}(y) = z'$ and $\tau_{Z,X}(z) = x$. Hence $\tau_{Y,Z'}h\tau_{Z,X}|_Z(z) = \tau_{Y,Z'}h(x) = \tau_{Y,Z'}(y) = z' = \gamma_{x,y}(z)$.
(4) *Global definition of $E(h)$.* Now an easy induction along geodesics in G_i which start in X shows that we can, in precisely one way, extend the definition of $E(h)$ to horoballs and elements of $\mathcal{C}_i$ in the neighborhood of $\tilde{Z}$ for any $\tilde{Z} \in \mathcal{C}_i$ on which $E(h)$ is already defined. We thus get an automorphism $E(h)$ of T, and it is clear by construction that $E(h)$ is level-preserving.
(5) *$E(h)$ is an element of L_i.* For any horoball $H = \mathrm{HB}(v)$, ($v \in VT$ with $\ell(v) = i$), there is by construction of $E(h)$ a $z_1 \in HS = \mathrm{HS}(v)$ such that $E(h)|_H =$

$\gamma_{z_1,z_2}|_H$, where $z_2 = E(h)(z_1)$. But then by Lemma 49, $E(h)|_H = \gamma_{w_1,w_2}|_H$ with $w_2 = E(h)(w_1)$ for all $w_1 \in HS$, yielding Condition (a) of Definition 50. To check Condition (b), we first consider an edge $\{A, B\}$ in G_i such that A is nearer to X than B, and set $A' = E(h)(A)$, $B' = E(h)(B)$. By the inductive definition of $E(h)$, we have $E(h)\tau_{A,B}E(h)^{-1}|_{A'} = \tau_{A',B'}|_{A'}$. Similarly for all edges on the geodesic in G_i between X and B. Applying Definition 43 to $\tau_{X,B}$, we now obtain $E(h)\tau_{X,B}E(h)^{-1}|_Y = \tau_{Y,B'}|_Y$. Since $B \in \mathcal{C}_i$ can be chosen arbitrarily, Remark 51 (2) implies that Condition (b) is also satisfied. □

The next proposition shows that the above extension procedure is compatible with commensurators. This is [BM96, Proposition 8.6], which we reproduce to correct minor misprints and because we deal with the group Δ rather than Γ. We denote the fact that a subgroup N is of finite index in a group H by writing: $N <_{\text{f.i.}} H$. Since $\Delta <_{\text{f.i.}} \Gamma$, the commensurators $C(\Delta)$ and $C(\Gamma)$ of these groups in G coincide. We keep the notations $Y_i = C_i(x_0)$ and $\Delta_i = \mathrm{Stab}_\Delta(Y_i)$ introduced in Lemma 46 and set $G_i := \mathrm{Aut}(Y_i)$, $G_i' := \{h \in G_i | \ell(h \cdot x) = \ell(x) \text{ for all } x \in VY_i\}$.

Proposition 54. *Let $\bar{\Delta}_i$ be the natural image of Δ_i in G_i'. Then for any $g \in \mathrm{Comm}_{G_i'}(\bar{\Delta}_i)$, the unique extension $E(g) \in L_i$ provided by Proposition 53 lies in the commensurator* $\mathrm{Comm}_L(\Delta) = C(\Delta) \cap L$.

Proof. We first remark that the map $E : G_i' \to L_i$, $h \mapsto E(h)$ is a homomorphism by the uniqueness statement of Proposition 53. It is enough to show that for any $g \in \mathrm{Comm}_{G_i'}(\bar{\Delta}_i)$, we have $\Delta \cap E(g)\Delta E(g)^{-1} <_{\text{f.i.}} \Delta$. (Indeed, apply it to g^{-1} and conjugate by $E(g)$ to deduce that $\Delta \cap E(g)\Delta E(g)^{-1} <_{\text{f.i.}} E(g)\Delta E(g)^{-1}$.) By definition of a commensurator, there is $\{\delta_j\}_{j=1}^r \subset \Delta_i$ such that, with $\bar{\delta}_j := \delta_j|_{Y_i}$, $\bar{\Delta}_i = \bigsqcup_{j=1}^r \bar{\delta}_j(\bar{\Delta}_i \cap g\bar{\Delta}_i g^{-1})$. It is enough to prove that we have: $\Delta = \bigcup_{j=1}^r \delta_j(\Delta \cap E(g)\Delta E(g)^{-1})$. Let $\delta \in \Delta$. Since $\tau_{\delta \cdot Y_i, Y_i} \circ \delta$ stabilizes Y_i, there is an index j such that $\bar{\delta}_j^{-1} \circ (\tau_{\delta \cdot Y_i, Y_i} \circ \delta)|_{Y_i} \in \bar{\Delta}_i \cap g\bar{\Delta}_i g^{-1}$. In particular, $\sigma := g^{-1} \circ (\bar{\delta}_j^{-1} \circ \tau_{\delta \cdot Y_i, Y_i} \circ \delta)|_{Y_i} \circ g$ lies in $\bar{\Delta}_i$. Now we consider $E(g)^{-1}(\delta_j^{-1}\delta)E(g) \in L_i$ (recall that $\Delta \subseteq L_i$ by Claim 52) and prove that its restriction to Y_i coincides with the restriction of an element of Δ to Y_i. By uniqueness (Proposition 53), this will show that $E(g)^{-1}(\delta_j^{-1}\delta)E(g) \in \Delta$. The element δ being arbitrary in Δ, this will finally prove: $\Delta \cap E(g)\Delta E(g)^{-1} <_{\text{f.i.}} \Delta$. We are thus led to computing:

$$(E(g)^{-1}(\delta_j^{-1}\delta)E(g))|_{Y_i} = E(g)^{-1}|_{\delta_j^{-1}\delta \cdot Y_i} \circ \delta_j^{-1}|_{\delta \cdot Y_i} \circ \delta|_{Y_i} \circ g$$

$$= E(\bar{\delta}_j g)^{-1}|_{\delta \cdot Y_i} \circ \delta|_{Y_i} \circ g.$$

We make σ appear on the right to obtain an element of $\bar{\Delta}_i$:

$$= E(\bar{\delta}_j g)^{-1}|_{\delta \cdot Y_i} \circ ((\tau_{\delta \cdot Y_i, Y_i}|_{\delta \cdot Y_i})^{-1} \circ \bar{\delta}_j \circ g)$$

$$\circ (g^{-1} \circ \bar{\delta}_j^{-1} \circ \tau_{\delta \cdot Y_i, Y_i}|_{\delta \cdot Y_i}) \circ \delta|_{Y_i} \circ g$$

$$= E(\bar{\delta}_j g)^{-1}|_{\delta \cdot Y_i} \circ ((\tau_{\delta \cdot Y_i, Y_i}|_{\delta \cdot Y_i})^{-1} \circ \bar{\delta}_j \circ g) \circ \sigma$$

$$= (E(\bar{\delta}_j g)^{-1}|_{\delta \cdot Y_i} \circ \tau_{Y_i, \delta \cdot Y_i}|_{Y_i} \circ E(\bar{\delta}_j g)|_{Y_i}) \circ \sigma.$$

Since $E(\bar{\delta}_j g) \in L_i$, this last product is $\tau_{Y_i,Z}|_{Y_i} \circ \sigma$, where $Z := E(\bar{\delta}_j g)^{-1}(\delta \cdot Y_i)$. Because $\tau_{Y_i,Z} \in \Delta$ and $\sigma \in \bar{\Delta}_i$, the restriction of $E(g)^{-1}(\delta_j^{-1}\delta)E(g)$ to Y_i coincides with the restriction of an element of Δ to Y_i. □

4.2.5. *Density*

We are now in a position to complete the proof of our main density theorem. The main tools will be Proposition 54 and Liu's theorem on the arithmeticity of uniform tree lattices.

Proof of Theorem 4. Given any level-preserving isomorphism $\phi : T_1 \to T_2$ between two finite subtrees T_1 and T_2 of T, we have to find an element $h \in C(\Gamma) \cap L = C(\Delta) \cap L$ such that $h|_{T_1} = \phi$. We choose $i > 0$ big enough so that the subtree $Y_i = C_i(x_0)$ of T introduced in Lemma 46 satisfies the following two conditions:

(i) the subtree Y_i contains $T_1 \cup T_2$;
(ii) the tree Y_i is not biregular.

We briefly explain why Condition (ii) can always be achieved and why we need it. First recall that the vertices of level $j < i$ in Y_i have the same degree in Y_i as in T, namely $q_j + 1$ (Facts 12 and 13). And the vertices of level i are of degree q_i in Y_i. So either T is not biregular, and there exists an l such that $q_l \neq q_{l+2}$; then we choose $i > l + 2$. Or T is biregular, with degrees $q_0 + 1$ and $q_1 + 1$. In this case Y_i is not biregular for any $i \geq 2$. If Y_i is not biregular, then we know from Lemma 16, Remark 17 and Lemma 46 (i) that all automorphisms of Y_i are automatically level–preserving. This means $G_i' = G_i = \mathrm{Aut}(Y_i)$ in the notation of Proposition 54.

Now $\bar{\Delta}_i$, the image of Δ_i in G_i, is a uniform Y_i-lattice (Lemma 46 (iii)). So by Liu's theorem, the commensurator $\mathrm{Comm}_{G_i}(\bar{\Delta}_i)$ is dense in $G_i = G_i'$. Hence there exists a $g \in \mathrm{Comm}_{G_i'}(\bar{\Delta}_i)$ satisfying $g|_{T_1} = \phi$. By Proposition 53, g can be (uniquely) extended to an element $h := E(g)$ of L_i. Then Proposition 54 implies that $h \in C(\Delta) \cap L$, and by construction $h|_{T_1} = \phi$. □

Theorem 4 has some immediate consequences which are worth to be mentioned. The first one is the generalization of the example discussed in [BM96, Section 8.3] to the class of all lattices of directly split Nagao type with associated non-biregular tree T. Recall that $L = G$ in these cases (see Lemma 16).

Corollary 55. *If Γ is a T-lattice of directly split Nagao type and T is not biregular, then $C(\Gamma)$ is dense in $G = \mathrm{Aut}(T)$.* □

The next corollary combines Theorem 4 and Proposition 20. It was stated as Theorem 5 in the introduction.

Corollary 56. *If Γ is a T-lattice of directly split Nagao type and T is biregular, then either $C(\Gamma)$ is a subgroup of L which is then dense in L or else the closure of $C(\Gamma)$ in G contains G°.* □

For regular trees, this has the following immediate consequence.

Corollary 57. *If Γ is a T-lattice of directly split Nagao type such that T is regular and $C(\Gamma)$ contains a type-interchanging automorphism, then $C(\Gamma)$ is dense in G.*

If T is biregular, it is in general hard to see (and we do not know of any general method how to decide this question) whether $C(\Gamma)$ is included in L or not. However, there is one class of lattices of Nagao type where $C(\Gamma) \not\subseteq L$ is obvious, and this class of lattices we are going to discuss in the next section.

4.3. Applications to Moufang twin trees

We introduce Moufang twin trees and recall some decompositions available for groups acting on these trees: this enables us to connect this theory to the previous section. We also provide examples, from the Nagao lattice to more exotic twin trees, including generalized rank 2 Kac-Moody groups. General references for twin trees are [RT94], [RT99] and [Tit90, §9].

4.3.1. *Twin trees without groups*

The notion of codistance relating two trees can be introduced without reference to any group action [RT94, §1]. In this first subsection, we introduce as many notions as possible purely in combinatorial terms. We will show later how they coincide with previously defined objects in the Nagao context, when we have enough automorphisms.

Definition 58. *Let $(T_\pm, \delta^*)$ be a triple where T_+ and T_- are trees and where δ^* is a function with values in $\mathbf{N}$ on pairs of vertices of opposite signs in $T_+ \sqcup T_-$. We say that $(T_\pm, \delta^*)$ is a* twin tree *or, equivalently, that δ^* is a* codistance *between T_+ and T_- if δ^* satisfies:*

(Codist) *For any $x_+ \in VT_+$ and any $y_- \in VT_-$, we have $\delta^*(y_-, x_+) = \delta^*(x_+, y_-)$, and furthermore, setting $m := \delta^*(x_+, y_-)$: for each $y'_- \in VT_-$ adjacent to y_-, one obtains $\delta^*(x_+, y'_-) = m \pm 1$; if $m > 0$ there is a unique y'_- such that $\delta^*(x_+, y'_-) = m + 1$; and we require the similar conditions with T_+ and T_- interchanged.*

In this case, we say that two vertices of opposite signs are opposite *if their codistance is* 0.

An automorphism *of the twin tree $(T_\pm, \delta^*)$ is a pair (α_+, α_-) of automorphisms of T_+, respectively T_-, such that $\delta^*(\alpha_+(x_+), \alpha_-(y_-)) = \delta^*(x_+, y_-)$ for any $x_+ \in VT_+$ and any $y_- \in VT_-$. We denote the group of all automorphisms of $(T_\pm, \delta^*)$ by A.*

Two vertices in a tree have the same type if they are at even distance from one another. Adding that two vertices of opposite sign have the same type if they are at even codistance, we obtain an equivalence relation (with two classes) on the vertices of $T_+ \sqcup T_-$ [RT94, p. 406]. Two thick trees (i.e. two trees where all vertices have degree at least 3) related by a codistance are isomorphic and biregular because any two vertices of the same type have the same degree [RT94, Proposition 1]. We denote the types by 0 and 1 and write them as subscripts when needed. Note that we do not require automorphisms of twin trees to be type-preserving. We denote by A° the subgroup of A of all type-preserving automorphisms of $(T_\pm, \delta^*)$. Obviously, $[A : A^\circ] \leq 2$.

Let x_0^+ and x_1^+ be two adjacent vertices in T_+. We can choose two vertices x_0^- and x_1^- such that x_i^+ and x_i^- are opposite, $i \in \{0; 1\}$: these vertices will be called the *standard positive* or *negative* vertices of type 0 or 1, accordingly. A pair of geodesic lines $L_\pm \subset T_\pm$ is called a *twin apartment* if any vertex in $L := L_+ \cup L_-$ has a unique opposite in L. By [RT94, Proposition 3.5] there is a unique twin apartment containing the four standard vertices, which we call the *standard twin apartment*. It follows from the axioms of a codistance that the codistance between to vertices of opposite sign in a twin apartment is the distance between any of the two points to the unique opposite of the other one in the twin apartment.

A straightforward consequence of (Codist) is a monotonicity property distinguishing a subfamily of geodesic rays and boundary points in each tree [RT94, 3.1–3.4]: for each pair of non-opposite vertices x_+ and y_- there is a unique ray in T_+ (resp. in T_-) emanating from x_+ (resp. y_-) along which the codistance from y_- (resp. x_+) is increasing. This defines a unique point $\xi(x_+)_{y_-}$ of the ideal boundary $\partial_\infty T_+$ and a unique point $\eta(y_-)_{x_+}$ of the ideal boundary $\partial_\infty T_-$, so that the rays are $[x_+; \xi(x_+)_{y_-})$ and $[y_-; \eta(y_-)_{x_+})$. Using a standard pair of opposite vertices $x_i^\pm$, we see that the above monotonicity arguments naturally define in $T_\pm$ a union of geodesic rays, which we call the *standard infinite star of type i* and which we denote by $\mathrm{St}^\infty(x_i^\pm)$. The number of rays from $x_i^\pm$ in $\mathrm{St}^\infty(x_i^\pm)$ is the valency of this vertex.

4.3.2. *Root groups and Moufang condition*

The root system Φ attached to an arbitrary Coxeter system is defined in [Tit87, Sect. 5] or in [Hum90, II.5]. We are interested in the specific case where the Coxeter group is the infinite dihedral group $D_\infty = \mathbf{Z}/2 * \mathbf{Z}/2$, say with generators s_0 and s_1. The associated Coxeter complex is the tiling of the real line $\mathbf{R}$ by the segments $[n; n+1]$, $n \in \mathbf{Z}$. Let s_0 (resp. s_1) be the reflection $x \mapsto -x$ (resp. $x \mapsto 2-x$). The *roots* of Φ are the half-lines defined by the integers, the positive ones being those containing $[0, 1]$. Each vertex has type 0 or 1, the boundary of a root is called its *vertex*, and the *type* of a root a is the type of its vertex. Two roots a and b are *prenilpotent* with one another if $a \subseteq b$ or $b \subseteq a$ [Tit87]. Let $a_0 := [0; +\infty)$ and $a_1 := (-\infty; 1]$ be the roots whose intersection is the edge $E := [0; 1]$. A useful viewpoint on twin apartments is to see them as two Tits cones of the same Weyl group glued along their tip [Rem02b, 5.3.2]. In our case, a twin apartment is the double cone generated by the above tiling of the real line. We recover the geodesics $L_\pm$ as the affinizations of the double cone. A *twin root* generated by a root a of any sign is then the half-plane bounded by the line passing through the origin and the vertex of a, and containing a. We denote it by a, too. Two twin roots generated by prenilpotent roots a and b are also called prenilpotent.

For a single building, the *Moufang condition* is discussed in detail in [Ron89, §6]. For twin trees it is introduced in [RT94, pp. 475–476] as follows. Let a be a twin root in the twin apartment L. Let us denote by U_a the group of automorphisms of $(T_\pm, \delta^*)$ fixing the half twin apartment a and every edge having a vertex in the interior of a.

By [RT94, Proposition 4.1], the group U_a acts freely on the set of twin apartments containing a, but it may be trivial. If the action is transitive we say that U_a is a *root group*.

Definition 59. *We say that a twin tree* $(T_{\pm}, \delta^*)$ *satisfies the* Moufang property *or is a* Moufang twin tree *if there is a twin apartment L in which U_a is a root group for each twin root $a \subset L$. We denote by Λ the subgroup of A° generated by the root groups U_a.*

The definition seems to depend on the choice of a twin apartment, but if a twin tree is Moufang, then Λ is transitive on the set of twin apartments [RT94, Proposition 4.5]. Note that so far the roles of the trees T_+ and T_- are symmetric. The Moufang property says that the valency at a vertex v is 1 greater than the order of the root group attached to a twin root whose boundary contains v, so the trees are locally finite if and only if the root groups are finite, in which case the pointwise fixator of a twin apartment, which we denote by H, is finite too. We now introduce a subgroup of Λ for which the symmetry doesn't hold any longer.

Definition 60. *We denote by* $\mathrm{Stab}_\Lambda(x_\epsilon)$ *(where $\epsilon \in \{+, -\}$) the stabilizer in Λ of any point $x_\epsilon \in T_\epsilon$. We use Γ instead of* $\mathrm{Stab}_\Lambda(x_\epsilon)$ *when x_ϵ is the vertex x_0^- or x_1^- and the choice of one of these two points is clear.*

The group $\mathrm{Stab}_\Lambda(x_\epsilon)$ naturally acts on the tree of sign $-\epsilon$. Since Λ acts edge-transitively on T_-, up to conjugacy the possible Γ-actions are those of $\mathrm{Stab}_\Lambda(x_0^-)$, of $\mathrm{Stab}_\Lambda(x_1^-)$ and of their intersection $\mathrm{Stab}_\Lambda([x_0^-; x_1^-])$. Henceforth, we fix a vertex $v_- := x_i^-$, $i \in \{0; 1\}$, and consider the corresponding Γ-action on the positive tree T_+. The subgroup generated by the root groups U_a indexed by the twin roots a containing v_- is a finite index subgroup of $\Gamma = \Lambda(v_-)$.

Picture 61.

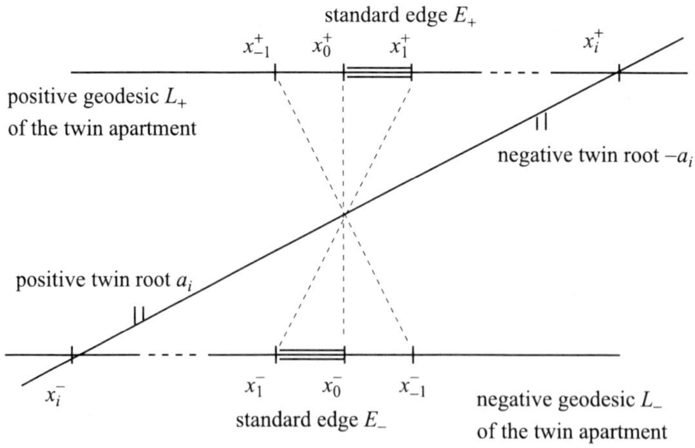

4.3.3. *Connection with lattices of Nagao type*

Let us first recall the technical condition on Moufang twin trees already stated in the introduction.

Definition 62. *We say that a Moufang twin tree* $(T_\pm, \delta^*)$, *or the associated group* Λ, *satisfies condition* (Comm) *if any two root groups* U_a *and* U_b *commute whenever* a *and* b *are different, prenilpotent twin roots.*

We can now formulate the result relating Moufang twin trees and lattices of Nagao type.

Proposition 63. *Let* $(T_\pm, \delta^*)$ *be a thick locally finite Moufang twin tree. Let* A *be its automorphism group and let* Λ *be the subgroup generated by the root groups. We fix a twin apartment* $L_\pm$, *a pair of opposite vertices* $v_\pm \in L_\pm$, *and we denote by* Γ *the group* $\mathrm{Stab}_\Lambda(v_-)$. *For each sign* $\epsilon = \pm$, *we number* $\{v_j^\epsilon\}_{j\in\mathbf{Z}}$ *the vertices of* L_ϵ *in such a way that* $v_0^\epsilon = v_\epsilon$ *and for any* j, v_j^ϵ *and* v_{j+1}^ϵ *are neighbours and* v_j^ϵ *and* $v_j^{-\epsilon}$ *are opposite. For each* $j \in \mathbf{Z}$, *we denote by* V_j *(resp.* U_j*) the positive (resp. negative) root group associated to the twin root whose boundary is* $\{v_j^+; v_j^-\}$, *and we denote by* E_+ *(resp.* E_-*) the edge* $[v_0^+; v_1^+]$ *(resp.* $[v_0^-; v_1^-]$*).*

(i) *Any of the two geodesic rays* $\{v_j\}_{j\geq 0}$ *and* $\{v_j\}_{j\leq 0}$ *from* v_+ *is a fundamental domain for the* Γ*-action on* T_+.

(ii) *The stabilizer* $\mathrm{Stab}_A(E_+ \cup \{v_-\})$ *is finite, and in particular so is* $H_0 := \mathrm{Stab}_\Gamma(E_+)$. *Moreover we have the decomposition:* $\mathrm{Stab}_\Gamma(v_j) = H_0 \ltimes (U_1, \ldots, U_j)$ *for each* $j > 0$, *and the stabilizer* $\mathrm{Stab}_\Gamma(v_0)$ *is the finite group generated by* H_0 *and the root groups indexed by the two opposite twin roots bounded by* v_- *and* v_+.

(iii) *The group* Γ, *as well as its action on* T_+, *identifies with the group and the action attached to the graph of groups of Picture 4.1, where the geodesic ray is* $\{v_j\}_{j\geq 0}$ *and the groups are the above stabilizers.*

(iv) *The group* Γ *is a* T_+*-lattice of Nagao type, which is of directly split Nagao type whenever* $(T_\pm, \delta^*)$ *satisfies* (Comm).

(v) *The level function on the positive vertices is nothing else than the codistance from the* Γ*-fixed negative vertex* v_-, *that is:* $\ell(x) = \delta^*(v_-, x)$ *for any* $x \in VT_+$.

Proof. Recall that two edges are called *opposite* if the two corresponding pairs of vertices of opposite signs and same type are pairs of opposite vertices. According to [Tit92, Proposition 7], the groups A and Λ both admit a structure of RGD-system [Tit92, 3.3], also called a twin root datum – see also [Abr97, I, Definition 2] or [Rem02b, Chapter 1]. To the RGD-system is naturally attached a positive (resp. negative) Tits system whose Borel subgroups are the stabilizers of the edges in T_+ (resp. in T_-) [Abr97, §1, Proposition 1]. The group Γ is a parabolic subgroup of the negative Tits system and (i) is a special case of [Abr97, §3, Corollary 1], which applies to general Moufang twin buildings.

We have $\delta^*(v_1^+, v_0^-) = 1$ because v_1^+ and v_1^- are opposite and v_1^- and v_0^- are neighbors. Since $L_\pm$ is a twin apartment, v_1^- is the only neighbor of v_0^- in L_- which is opposite v_1^+, so that v_{-1}^- is the only neighbor of v_0^- to be at codistance 2 from v_1^+. Therefore the positive root group V_0 acts transitively on the neighbors of v_0^- which are $\neq v_{-1}^-$. Let $h \in \mathrm{Stab}_A(E_+ \cup \{v_-\})$; by the previous sentence, we can find $u \in V_0$ such that $u^{-1}h \in \mathrm{Stab}_A(E_+ \cup E_-)$, so that $\mathrm{Stab}_A(E_+ \cup \{v_-\}) = V_0 \cdot \mathrm{Stab}_A(E_+ \cup E_-)$.

Finally, the finiteness of $\mathrm{Stab}_A(E_+ \cup E_-)$ follows from Ronan–Tits' rigidity theorem [RT94, Theorem 4.1], which says that the identity is the only twinning automorphism which fixes E_+, E_- and all the edges having a vertex in common with E_+. Since $\Lambda < A$, this obviously implies the finiteness of $H_0 = \mathrm{Stab}_\Gamma(E_+)$. The rest of (ii) is a special case of Levi decompositions of stabilizers of pairs of points of opposite signs in twin buildings [Rem02b, 6.3.4].

For a general Moufang twin tree, the group $U_1, \ldots, U_i$ is in bijection with the set $U_1 \times \cdots \times U_i$, and it is isomorphic to the direct product group $U_1 \times \cdots \times U_i$ whenever (Comm) is satisfied: this proves (iv), once we note that (iii) is a classical consequence of Bass–Serre theory [Ser77, I.4.5, Théorème 10]. At last, (v) follows from the fact that the two functions $\ell(\cdot)$ and $\delta^*(v_-, \cdot)$ coincide on the ray $\{v_j\}_{j\geq 0}$ and are constant on each Γ-orbit. □

Remark 64. In terms of twin root data (or, equivalently, of RGD-systems), the group Δ of 4.2.1 is the unipotent radical of the negative parabolic subgroup Γ, and the decomposition $\Gamma = \Gamma_0 \ltimes \Delta$ is a Levi decomposition [Rem02b, 6.2.2]. The fundamental domain F of Lemma 35 is the union of the level-increasing rays from v_+: it is the infinite star $\mathrm{St}^\infty(v_+)$ defined purely in terms of codistance in 4.3.1.

Recall that two subgroups of a given group are *commensurable* if they share a finite index subgroup.

Lemma 65. *With the same notation as above, we have:*

(i) *For any two points $x, x' \in T_-$, the groups $\mathrm{Stab}_A(x)$ and $\mathrm{Stab}_A(x')$ are commensurable.*
(ii) *For any point $x \in T_-$, we have $A = \mathrm{Aut}(T_\pm, \delta^*) < C(\Gamma)$.*
(iii) *The group $C(\Gamma)$ acts transitively on geodesics of given length and type in T_+, and it acts transitively on geodesics of given length if T_+ has a type-exchanging automorphism.*

Proof. For any $x \in T_-$, we can choose an edge E whose closure contains v, and in this case we have: $\mathrm{Stab}_{A^\circ}(E) <_{\text{f.i.}} \mathrm{Stab}_A(x)$. Therefore, in order to prove (i) it is enough to consider the case when x and x' both belong to the interior of an edge, say E for x and E' for x'. Let y (resp. y') be the midpoint of E (resp. E'), and let δ be the distance between y and y'. Let g be an arbitrary automorphism of T_+ stabilizing the edge E. Then $g \cdot y'$ is at distance δ from y and when g is type-preserving, the geodesic from y to $g \cdot y'$ leaves E through the vertex through which the geodesic from y to y' leaves E. Using root groups for roots containing E, we see for any $g \in \mathrm{Stab}_{G^\circ}(E)$, the edge $g \cdot E'$ is a $\mathrm{Stab}_\Lambda(E)$-transform of E'. Since T_- is locally finite, we have a finite set of such edges, say $h_1 \cdot E', \ldots, h_N \cdot E'$ with $h_1, \ldots, h_N \in \mathrm{Stab}_\Lambda(E)$. Therefore $\mathrm{Stab}_{G^\circ}(E) = \bigsqcup_{j=1}^m h_j(\mathrm{Stab}_{G^\circ}(E) \cap \mathrm{Stab}_{G^\circ}(E'))$, so that $\mathrm{Stab}_{A^\circ}(E) \cap \mathrm{Stab}_{A^\circ}(E') <_{\text{f.i.}} \mathrm{Stab}_{A^\circ}(E)$. Switching x and x' and using the first remark, we obtain (i). Moreover (ii) follows from (i) since $g\mathrm{Stab}_G(x)g^{-1} = \mathrm{Stab}_G(g \cdot x)$ for any $g \in G$. The subgroup Λ already enjoys the transitivity property of the first assertion of (iii) (this is true in the general twin building case [Abr97, §2, p. 28]), and the second assertion is easily deduced from the first one. □

We can finally turn to the proof of our second main result.

Proof of Theorem 7. We choose $(T_\pm, \delta^*)$ a thick locally finite Moufang twin tree satisfying (Comm), and go on using the notation of the previous two results. By Proposition 63 (iv), the lattice Γ is of directly split Nagao type. The closure group $\overline{C(\Gamma)}$ contains $\overline{C(\Gamma) \cap L}$, which is equal to L by Theorem 4. But $\overline{C(\Gamma)}$ also contains Λ, in which many elements do not preserve the level function ℓ. To see this, we note that by Proposition 63 (v) we have: $\ell(x) = \delta^*(v_-, x)$ and $\ell(g \cdot x) = \delta^*(g^{-1} \cdot v_-, x)$ for any $x \in VT_+$ and any $g \in A$. In order to have $\ell(x) \neq \ell(g \cdot x)$, it is enough to make $x := v_+$ and to pick an element $n \in N_\Lambda(L_\pm)$ lifting a reflection in a vertex x_j with $j \neq 0$ (which always exists e.g. according to [Abr97, §2, Lemma 4 (ii)]). Finally it remains to apply Proposition 20 to the group $F = \overline{C(\Gamma)}$, which contains $L \cup \Lambda$. □

4.3.4. *Examples*

Let us now give examples of Moufang twin trees from the most familiar to the most exotic ones. We first note that according to M. Ronan and J. Tits [RT99, Corollary 8.2], a thick semihomogeneous tree whose set of vertices has cardinality α, belongs to 2^α isomorphism classes of twinnings, among which most of them have no automorphism. The examples below are all required to be Moufang (in particular, have big automorphism groups).

Example 66. There is a detailed study of $\Lambda = \mathrm{PSL}_2(\mathbf{F}_q[t, t^{-1}])$ in [RT94, §2]. Let M be a rank 2 free $\mathbf{F}_q[t, t^{-1}]$-lattice with basis $\{u; v\}$, and let T_+ and T_- be the Bruhat–Tits trees of $\mathrm{SL}_2(\mathbf{F}_q((t)))$ and $\mathrm{SL}_2(\mathbf{F}_q((t^{-1})))$, respectively [Ser77, II.1]. The vertices in T_+, resp. in T_-, are the homothety classes of $\mathbf{F}_q[[t]]$-lattices, resp. of $\mathbf{F}_q[[t^{-1}]]$-lattices, in the vector space $M \otimes_{\mathbf{F}_q[t,t^{-1}]} \mathbf{F}_q((t))$, resp. in $M \otimes_{\mathbf{F}_q[t,t^{-1}]} \mathbf{F}_q((t^{-1}))$. The codistance is defined in [RT94, Lemma 2.1]. A useful subset of vertices in T_+ is given by the homothety classes of the lattices $\mathbf{F}_q[[t]]u \oplus \mathbf{F}_q[[t]]t^j v$ for $j \in \mathbf{Z}$, and similarly in T_- replacing t by t^{-1}. The convex hull of these vertices in T_+ is a geodesic L_+, and together with the similar convex hull L_- in T_-, they form a twin apartment L. The pointwise fixator H of $L_\pm$ is the finite subgroup of diagonal matrices with coefficients in $\mathbf{F}_q^\times$. The Nagao lattice $\Gamma = \mathrm{PSL}_2(\mathbf{F}_q[t^{-1}])$ lies in Λ and the latter group acts edge-transitively on each tree. It can actually be proved by hand that for each $d \in \mathbf{Z}$, the group $\left\{\left(\begin{smallmatrix} 1 & \lambda \cdot t^d \\ 0 & 1 \end{smallmatrix}\right) : \lambda \in \mathbf{F}_q\right\} \simeq (\mathbf{F}_q, +)$ is the root group attached to the root a_d defined as the convex hull of the homothety classes of the lattices $\mathbf{F}_q[[t]]u \oplus \mathbf{F}_q[[t]]t^j v$ for $j \geq d$. The union of these root groups when d ranges over $\mathbf{Z}$ is the unipotent radical of the upper triangular Borel subgroup, and condition (Comm) holds because this unipotent group is abelian.

Example 67. Split Kac-Moody groups of rank two: these groups are defined by generators and relations in full generality in [Tit87]. The datum needed to define such a group roughly consists in a generalized Cartan matrix A and a groundfield $\mathbf{K}$. The rank 2 condition says that the matrix is 2×2, the size 2 corresponding to the number of generators of the Weyl group. The diagonal entries are equal to 2, and one must require that the product of the off-diagonal entries be ≥ 4 so that the Weyl group of the building is infinite (dihedral). Each of the off-diagonal coefficients is negative, and the group satisfies the assumption (Comm) if and only if both entries are ≤ -2. This follows from the computation of commutator relations between root groups due, up to signs, to J. Morita [Mor88, §3, example 6]. Note that apart from the density of the commensurator for the rank 2 case, many other arguments supporting the analogy between (parabolic subgroups of) Kac-Moody groups over finite fields and arithmetic groups over function fields are given in [Rem02a].

Example 68. Twisted Kac-Moody groups are defined in [Rem02b, Part II, §11]. They are groups consisting of fixed points in a split Kac-Moody group, for a suitable Galois action. The rank 2 condition refers to the number of generators in the Weyl group. A classical example is provided by the unitary group $\mathrm{SU}_3(\mathbf{F}_q[t, t^{-1}])$, whose twinned trees are semihomogeneous of valencies $1 + q$ and $1 + q^3$ [Rem04, §3.5]. The unipotent radicals of Borel subgroups of SU(3) over $\mathbf{F}_q((t))$ or $\mathbf{F}_q((t^{-1}))$ are not abelian but only metabelian: $\mathrm{SU}_3(\mathbf{F}_q[t, t^{-1}])$ doesn't satisfy (Comm). The example of a semi-homogeneous Moufang twin tree whose corresponding group Λ satisfies (Comm) is given for instance in [Rem02b, 13.3]. The twisted group is a subgroup of rational points in a Kac-Moody group naturally acting on buildings whose apartments are tilings of the hyperbolic plane.

Example 69. A general construction of Moufang twin trees is given in [Tit90, §9]. Its starting point are two abstract root group data of rank 1, and the classification of the latter structure is known in the finite case. As Tits showed, starting with this data, it is always possible to construct a Moufang twin tree satisfying our Condition (Comm) and having as root groups (up to isomorphism) those occurring in the original root group data of rank 1. An easy way to produce non-classical locally finite Moufang twin trees by this method was pointed out by Sh. Mozes to the second author. Taking the affine group $\mathbf{F}_l^\times \ltimes \mathbf{F}_l$ (l is any prime power) acting on the affine line $\mathbf{F}_l$, we obtain what Tits calls a Moufang set and hence a root group data of rank 1. We take two copies of this, and then J. Tits' method provides homogeneous Moufang twin trees with root groups isomorphic to $\mathbf{F}_l^\times$. This cannot occur, at least if $l - 1$ is not itself a prime power, for Bruhat-Tits or Kac-Moody trees since in the latter cases the valencies are cardinalities of projective lines.

Therefore, for instance with $l = 7$, we obtain a non-uniform tree lattice of directly split Nagao type for a regular tree of valency 7. In the latter case, the root groups are all isomorphic to $\mathbf{F}_7^\times \simeq \mathbf{Z}/2 \times \mathbf{Z}/3$. Let us use the notation of the previous subsection, in particular the T_+-lattice Γ is the parabolic subgroup of Λ fixing a negative vertex v_- of $(T_\pm, \delta^*)$. The subgroup H of Γ generated by the root groups indexed by the negative roots containing a given geodesic subray of L_- emanating from v_-, is isomorphic to the direct sum of these root groups. Taking the 2-torsion (resp. 3-torsion) part of H and arguing as in [Rem02a, Theorem 5], we conclude that H, hence Γ, cannot be linear over any field. In other words, we have obtained a group inclusion $\Gamma < G$ where the tree-lattice Γ is non-linear but arithmetic in a generalized sense, whereas the classical Nagao lattice is obviously linear. Of course, the construction works with any prime power l such that $l - 1$ admits two different prime divisors.

Example 70. In J. Tits' construction of the previous example, the starting point is a pair of two root group data. In the case where each root group datum comes from a PSL_2-action on a projective line, a down-to-earth construction is made in [RR06, § 2]. This is enough to produce strictly more general Moufang twin trees than those coming from Kac-Moody groups since there may be two ground fields (one for each type of vertex). The choices of the characteristics of the ground fields can be made so that the argument of the second paragraph of the previous example enables to obtain again ⟨⟨non-linear but arithmetic⟩⟩ tree-lattices. At last, the concrete viewpoint allows to generalize the construction to the case of Moufang twin buildings with a Weyl group of arbitrarily large rank, leading to generalized Kac-Moody groups with strong non-linearity properties acting on two-dimensional buildings [RR06, Theorem 4.A].

References

[Abr97] P. Abramenko, Twin buildings and applications to S-arithmetic groups, Lecture Notes in Mathematics, vol. 1641, Springer (1997).

[BL01] H. Bass and A. Lubotzky, Tree lattices, Progress in Mathematics, vol. 176, Birkhäuser (2001).

[BM96] M. Burger and Sh. Mozes, CAT(-1)-spaces, divergence groups and their commensurators, *Journal of the American Mathematical Society*, **9** (1996) 57–93.

[Hag03] F. Haglund, Linearity and commensurability of uniform lattices of right-angled hyperbolic buildings, preprint University Paris sud, Orsay (2003).

[Hum90] J. E. Humphreys, Reflection Groups and Coxeter Groups, Cambridge studies in advanced mathematics, vol. 29, Cambridge University Press (1990).

[Lif03] L. Lifschitz, Arithmeticity of rank-1 lattices with dense commensurators in positive characteristic, *Journal of Algebra*, **261** (2003) 44–53.

[Liu94] Y. Liu, Density of the commensurability group of uniform tree lattices, *Journal of Algebra*, **165** (1994) 346–359.

[Mar91] G.A. Margulis, Discrete Subgroups of Semisimple Lie Groups, Ergebnisse der Mathematik und ihrer Grenzgebiete (3), vol. 17, Springer (1991).

[Mor88] J. Morita, Root strings with three or four real roots in Kac-Moody root systems, *Tôhoku Mathematical Journal*, **40** (1988) 645–650.

[Moz99] Sh. Mozes, Trees, lattices and commensurators, Algebra, K-Theory, Groups, and Education: On the Occasion of Hyman Bass's 65th Birthday (T. Y. Lam and A. R. Magid, eds.), Contemporary Mathematics, vol. 243, American Mathematical Society (1999) pp. 145–151.

[Rem02a] B. Remy, Classical and non-linearity properties of Kac-Moody groups, Rigidity in Dynamics and Geometry (Newton Institute, 2000) (M. Burger and A. Iozzi, eds.), Springer-Verlag (2002) pp. 391–406.

[Rem02b] B. Remy, Groupes de Kac-Moody déployés et presque déployés, Astérisque, vol. 277, Société Mathématique de France (2002).

[Rem04] B. Remy, Kac-Moody groups: split and relative theories. Lattices, Groups: Geometric and Combinatorial Aspects (Bielefeld 1999) (Th. Müller, ed.), London Math. Soc. Lecture Note Series, vol. 311 (2004) pp. 487–541.

[Ron89] M. Ronan, Lectures on Buildings, Perspectives in Mathematics, vol. 7, Academic Press (1989).

[RR06] B. Remy and M. Ronan, Topological groups of Kac-Moody type, right-angled twinnings and their lattices, *Comm. Math. Helv.*, **81** (2006) 191–219.

[RT94] M. Ronan and J. Tits, Twin trees I, *Inventiones Mathematicä*, **116** (1994) 463–479.

[RT99] M. Ronan and J. Tits, Twin trees II, Local structure and a universal construction, *Israel Journal of Mathematics*, **109** (1999) 349–377.

[Ser77] J.-P. Serre, Arbres, amalgames, SL_2, Astérisque, vol. 46, Société Mathématique de France (1977).

[Tit87] J. Tits, Uniqueness and presentation of Kac-Moody groups over fields, *Journal of Algebra*, **105** (1987) 542–573.

[Tit90] J. Tits, Théorie des groupes, Annuaire Collège de France (Résumé de cours) (1990) pp. 87–103.

[Tit92] J. Tits, Twin buildings and groups of Kac-Moody type, Groups, Combinatorics & Geometry (LMS Symposium on Groups and Combinatorics, Durham, July 1990) (M. Liebeck and J. Saxl, eds.), LMS Lecture Notes Series, vol. 165, Cambridge University Press (1992) pp. 249–286.

Proc. Inst. Wrk.shps. Int. Conf. – Essays in Geometric Group Theory No. 9, 2009, pp. 105–124.

Kac-Moody Groups as Discrete Groups

Bertrand Rémy

Institut Camille Jordan – UMR 5208 CNRS / Lyon 1, Université de Lyon 1 – Claude Bernard, 21 avenue Claude Bernard – Bâtiment Jean Braconnier, 69622 Villeurbanne Cedex – France
e-mail: remy@math.univ-lyon1.fr

Abstract. This survey paper presents the discrete group viewpoint on Kac-Moody groups. Over finite fields, the latter groups are finitely generated; they act on new buildings enjoying remarkable negative curvature properties. The study of these groups is shared between proving results supporting the analogy with some S-arithmetic groups, and exhibiting properties showing that they are new groups.

Keywords. Kac-Moody group, Tits system, pro-p group, lattice, arithmetic groups, algebraic group, Bruhat-Tits building, hyperbolic building, commensurator superrigidity, linearity.

Subject Classification: 22F50, 22E20, 51E24, 53C24, 22E40, 17B67

Introduction

Kac-Moody groups were initially designed to generalize algebraic groups [Tit87]. They share many properties with the latter groups, mainly of combinatorial nature. For instance, they admit BN-pair structures [Bou81, IV.2], and actually a much finer combinatorial structure – called *twin root datum* – formalizing the existence of root subgroups permuted by a (possibly infinite) Coxeter group [Tit92] (survey papers on this are for instance [Tit89] and [Rem04a]). Moreover Kac-Moody groups and some twisted versions of them are expected to be the group side in the classification of a reasonable class of buildings, the so-called Moufang 2-spherical twin buildings. Our intention is not to go into detail about this; nevertheless, we briefly recall these facts in order to emphasize that the goal of this paper is to present a significant change of viewpoint on Kac-Moody groups.

The geometric counterpart to the group combinatorics of a BN-pair is the existence of an action on a remarkable geometry: a building. To any Coxeter system (W, S) is attached a simplicial complex Σ on the maximal simplexes of which the group W acts simply transitively [Ron89, §2]. A *building* is a simplicial complex, covered by subcomplexes all isomorphic to the same Σ – called *apartments*, and satisfying remarkable incidence properties: any two *facets* (i.e. simplices) are contained in an apartment, and given any two apartments $A, A' \simeq \Sigma$, there is a simplicial isomorphism $A \simeq A'$ fixing $A \cap A'$ – see [Bro89, p. 77], and also [Ron89, §3] for the chamber system approach.

In order to define a general Kac-Moody group Λ, we need an integral matrix A satisfying properties much weaker than those defining Cartan matrices in the classical sense of complex semisimple Lie algebras [Tit87, Introduction]. We also need to choose a ground field (which will always be a finite field $\mathbf{F}_q$ in what follows). Writing down a presentation of Λ would require a lot of combinatorial and Lie algebra material we wouldn't use later – see [Tit87, Subsect. 3.6] and [Rem02b, Sect. 9] for details. It is a basic result of the theory that to a Kac-Moody group is naturally attached a pair of twin buildings (via BN-pairs) on the product of which it acts diagonally (§5.1.2 of the present paper). The standard example of such a group is $\Lambda = \mathbf{G}(\mathbf{K}[t, t^{-1}])$ for $\mathbf{G}$ a semisimple group over a field $\mathbf{K}$. Now, if we choose the ground field to be $\mathbf{F}_q$, the latter example is an arithmetic group in positive characteristic, and in this case the buildings alluded to above are the Bruhat-Tits buildings of the non Archimedean semisimple Lie groups $\mathbf{G}(\mathbf{F}_q((t)))$ and $\mathbf{G}(\mathbf{F}_q((t^{-1})))$.

The latter examples are a very special case of Kac-Moody groups, called of *affine type*, but many other cases are available. On the one hand, this suggests one to see Kac-Moody groups as generalized arithmetic groups over function fields, which leads to natural questions, e.g. asking whether some classical properties of discrete subgroups of Lie groups are relevant or true. On the other hand, it can be shown that some new buildings can be produced thanks to Kac-Moody groups, and this leads to asking whether the groups attached to exotic buildings are themselves new. Note that «new» in this context means that the buildings are neither of spherical nor of affine type, i.e. don't come from the classical Borel-Tits (resp. Bruhat-Tits) theory on algebraic groups over arbitrary (resp. local) fields.

From the point of view of metric spaces, Kac-Moody theory is an algebraic way to construct spaces with non-positively curved, often hyperbolic, distances and admitting highly transitive isometry groups. The algebraic origin of these groups enables to obtain interesting structure results for various isometry groups (discrete or much bigger). Therefore, in the case of hyperbolic Kac-Moody buildings, techniques from group combinatorics such as BN-pairs and from hyperbolic spaces «à la Gromov» can be combined. This leads us to say that the general trend to study finitely generated Kac-Moody groups is from algebraic and combinatorial methods to geometric and dynamical ones. In this paper, we explain for instance how the theory of finitely generated Kac-Moody groups, i.e. Kac-Moody groups over finite fields, naturally leads to studying uncountable totally disconnected groups generalizing semsimple groups over local fields of positive characteristic, groups which we call *topological Kac-Moody groups*. We are especially interested in proving that the groups we obtain are new in general, by proving that they cannot be linear over any field. In short, the study of finitely generated Kac-Moody groups is shared between proving classical properties by comparing them to (linear) lattices of non Archimedean Lie groups, and finally standing by a difference with the classical situation to disprove linearities.

This approach is not new, since for lattices of products of trees M. Burger and Sh. Mozes managed to prove many classical properties of lattices (among which the normal subgroup theorem) but proved that an important difference is the possibility to obtain non residually finite groups, which is impossible for finitely generated linear groups [BM97], [BM00a], [BM00b]. The main application is the construction of the first finitely generated torsion free simple groups. We note also that Y. Shalom's work [Sha00] shows that thanks to representation theory, many properties can be proved for irreducible lattices of products of general locally compact groups. Since trees are one-dimensional special cases of buildings, the previous references encourage us to think that finitely generated Kac-Moody groups will produce interesting examples of groups which are not linear but handleable via their diagonal actions on products of buildings.

This paper is organized as follows. In Sect. 5.1, it is shown why Kac-Moody groups over finite fields should be seen as generalizations of some S-arithmetic groups in positive characteristic. It is also explained how they provide new buildings and why Kac-Moody groups should be expected to be new groups. In Sect. 5.2, we are interested in a specific class of hyperbolic buildings. In this context, we can produce non-isomorphic Kac-Moody groups with the same buildings, and discrete groups which are close to Kac-Moody groups, but with several ground fields: they have strong non-linearity properties. In Sect. 5.3, we are interested in totally disconnected groups generalizing semisimple groups over local fields, arising as closures of non-discrete actions of Kac-Moody groups on buildings. We quote the existence of a nice combinatorial structure, as well as a topological simplicity result for them. In Sect. 5.4, we sketch the proof of complete non-linearity of some Kac-Moody groups. This is where we use the topological groups of the previous section. We actually mention that there are some Kac-Moody groups all of whose linear images are finite, whatever the target field. In Sect. 5.5, we ask some questions about the various groups previously defined in the paper. We conjecture the non-linearity of a wide class of finitely generated Kac-Moody groups and the abstract simplicity of topological Kac-Moody groups.

The author expresses his deep gratitude to the organizers of the conference «Geometric Group Theory» (Guwahati, Assam – India), organized by the Indian Institute of Technology Guwahati and the Indian Statistical Institute, and supported by the National Board of Higher Mathematics. Meenaxi Bhattacharjee managed to make a tremendous human and scientific event out of a mathematical conference.

5.1. Generalized arithmetic groups acting on new buildings

This section is mainly dedicated to quoting results supporting the analogy between Kac-Moody groups over finite fields and $\{0; \infty\}$-arithmetic groups over function fields [Rem02a, Sect. 2–3]. The arguments are: the existence of a discrete diagonal action on a product of two buildings (5.1.1), cohomological finiteness properties (5.1.2) and continuous cohomology vanishings (5.1.3). In the last two subsections (5.1.4 and 5.1.5), we provide arguments showing that Kac-Moody theory does provide interesting new group-theoretic/geometric situations.

5.1.1. *Discrete actions on buildings and finite covolume*

A Kac-Moody group is generated by those of its root groups which are indexed by simple roots and their opposites (in finite number), and by a suitable maximal torus normalizing them [Tit87, §3.6]. Since over $\mathbf{F}_q$ all these groups are finite, we obtain:

Fact 5.1.1. Any Kac-Moody group Λ over any finite field is finitely generated.

From now on, Λ is a Kac-Moody group defined over $\mathbf{F}_q$. Recall that a group action on a building is *strongly transitive* if it is transitive on the inclusions of a chamber in an apartment. Combining [Tit87, Subsect. 5.8, Proposition 4] and [Ron89, Theorem 5.2], we obtain:

Fact 5.1.2. To Λ are attached two isomorphic, locally finite buildings $X_\pm$, each of them admitting a strongly transitive Λ-action.

This fact is fundamental to understanding Kac-Moody groups: the geometry of the buildings is the basic substitute for a natural structure on Λ arising from infinite-dimensional algebraic geometry. In the specific case of an S-arithmetic group $\Lambda = \mathbf{G}(\mathbf{F}_q[t, t^{-1}])$ (for $S = \{0; \infty\}$), the buildings $X_\pm$ are the Bruhat-Tits buildings of $\mathbf{G}(\mathbf{F}_q((t)))$ and $\mathbf{G}(\mathbf{F}_q((t^{-1})))$. In general, it is still true that for an arbitrary Kac-Moody group, the diagonal action on the product of buildings $X_- \times X_+$ is discrete [Rem99]. Moreover the action has a nice convex fundamental domain contained in a single apartment and defined as an intersection of roots (seen as half-apartments) [Abr97, §3, Corollary 1]. The next step in the analogy with arithmetic groups consists in asking whether the finitely generated group Λ is a *lattice* of $X \times X_-$, meaning that the locally compact group $\mathrm{Aut}(X_-) \times \mathrm{Aut}(X_+)$ moded out by the image of Λ carries a finite invariant measure [Mar91, 0.40]. This is the main result of [CG03] and [Rem99]:

Theorem 5.1.3. *Assume the Weyl group W of Λ is infinite and denote by $W(t) = \sum_{w \in W} t^{\ell(w)}$ its growth series. Assume that $W\left(\frac{1}{q}\right) < \infty$. Then Λ is a lattice of $X \times X_-$ for its diagonal action, and for any point $x_- \in X_-$ the stabilizer $\Lambda(x_-)$ is a lattice of X. These lattices are never cocompact.*

In the arithmetic case of $\mathbf{G}(\mathbf{F}_q[t, t^{-1}])$, the Weyl group has polynomial growth (it is virtually abelian because it is an affine reflection group), so the condition $W\left(\frac{1}{q}\right) < \infty$ is empty and $\mathbf{G}(\mathbf{F}_q[t, t^{-1}])$ is always a lattice of $\mathbf{G}(\mathbf{F}_q((t))) \times \mathbf{G}(\mathbf{F}_q((t^{-1})))$, whatever the value of the prime power q. This is a particular case of a well-known result in reduction theory in positive characteristic [Beh69], [Har69].

5.1.2. *Finiteness properties*

Cohomological finiteness is a very hard problem for arithmetic groups in positive characteristic [Beh03], which makes a sharp difference with the number field case. Nevertheless, it is natural to expect that some results, similar to those which are known for arithmetic groups in the function field case, should hold for Kac-Moody groups. This is indeed the case, up to taking into account more carefully the submatrices in the generalized Cartan matrix defining the group – this is the theme of P. Abramenko's book [Abr97]. The following theorem sums up Theorems 1 and 2 from [Abr03] in a slightly different language. For instance, we introduce the parabolic subgroups (defined by means of BN-pairs in [loc. cit.]) via the group action on the buildings: such a subgroup is a facet fixator. Note also that the below quoted results are still valid in the more general context of abstract *twin BN-pairs* [Abr97, §1].

Theorem 5.1.4. *Let Λ be a Kac-Moody group over $\mathbf{F}_q$, and let Γ be a facet fixator. Let Σ be the Coxeter complex of the Weyl group W, i.e. the model for any apartment in the building $X_\pm$ of Λ. Let R be a chamber of Σ and let Π be the set of reflections in the codimension one faces of R.*

(i) *If any two reflections of Π generate a finite group and if $q > 3$, then Γ is finitely generated; but it is not of cohomological type FP_2, in particular not finitely presented, whenever some set of three reflections in Π generates an infinite subgroup of W.*
(ii) *If any three reflections of Π generate a finite group and if $q > 6$, then Γ is finitely presentable; but it is not of cohomological type FP_3 whenever some set of four reflections in Π generates an infinite subgroup of W.*

When any two reflections of Π generate a finite group, the Kac-Moody group is called 2-*spherical*. Along with the Moufang property [Ron89, 6.4] and twinnings [Abr97, Definition 3], this notion plays a major role in the classification of buildings with infinite Weyl groups. In [Abr97] some further results are available; they deal with the higher cohomological finiteness properties FP_n and F_n [Bro87, VIII.5]. According to §5.1.1, finite presentability is the first finiteness property to be considered for the group Λ itself. A result due to P. Abramenko and B. Mühlherr [AM97] is available:

Theorem 5.1.5. *With the same notation as above, if any two reflections of Π generate a finite group and if $q > 3$, then Λ is finitely presentable.*

For instance, finite presentability holds for Kac-Moody groups whose buildings have chambers isomorphic to hyperbolic regular triangles of angle $\frac{\pi}{4}$ and $q > 3$, but never holds for those whose buildings are covered by chambers isomorphic to a regular right-angled r-gon, $r \geq 5$ (we will see in 5.1.4 that such Kac-Moody groups exist in both cases).

5.1.3. *Continuous cohomology and Kazhdan's property (T)*

By results of J. Dymara and T. Januszkiewicz, being 2-spherical also implies useful continuous cohomology vanishing for automorphism groups of buildings. The result below is a special case of [DJ02, Theorem E].

Theorem 5.1.6. *Let Λ be a Kac-Moody group over $\mathbf{F}_q$, defined by a generalized Cartan matrix A of size $n \times n$. Let $m < n$ be an integer such that all the principal submatrices of size $m \times m$ of A are Cartan matrices (i.e. of finite type). Then for $1 \leq k \leq m - 1$ and $q \gg 1$, the continuous cohomology groups $\mathrm{H}^k_{\mathrm{ct}}(\mathrm{Aut}(X_\pm), \rho)$ vanish for any unitary representation ρ.*

The first cohomology case is extremely useful since vanishing of $\mathrm{H}^1_{\mathrm{ct}}(G, \rho)$ for any unitary representation ρ is equivalent to property (T) [dlHV89, Chap. 4]. Therefore, when Λ is 2-spherical, Theorem 5.1.6 implies property (T) for the full automorphism groups $\mathrm{Aut}(X_\pm)$ with $q \gg 1$, hence for their product, and finally for any lattice in this product, by S.P. Wang's Theorem [Mar91, III, Theorem 2.12]. The above result says in particular that many Kac-Moody groups have property (T), a fundamental property for lattices of higher-rank simple algebraic groups [Mar91, III].

5.1.4. *Hyperbolic examples*

The existence of buildings with prescribed shapes of apartments and links around vertices is well-known in many cases [Ron86], [Bou97]. Some examples lead to interesting full automorphism groups and lattices [BP00], and some other examples have surprisingly few automorphisms, though they are quite familiar since they are Euclidean and tiled by regular triangles [Bar00]. The result below [Rem02c, Proposition 2.3] shows that some of these buildings are both relevant to Kac-Moody theory and to (generalized) hyperbolic geometry [BH99, III.H]. This enables to mix arguments of algebraic and geometric nature in the study of the corresponding Kac-Moody groups.

Proposition 5.1.7. *Let P be a polyhedron in the hyperbolic space $\mathbb{H}^n$, with dihedral angles equal to* 0 *or to $\frac{\pi}{m}$ with $m = 2, 3, 4$ or 6. For any prime power q, there is a Kac-Moody group whose building has constant thickness equal to $q+1$ and where the apartments are all tilings of $\mathbb{H}^n$ by P; this building is a complete geodesic $CAT(-1)$-metric space.*

Recall that the *thickness* at a codimension 1 cell C of a building is the number of maximal cells containing C. Note that according to G. Moussong [Mou88], any Coxeter group acts discretely and cocompactly on a $CAT(0)$-space, which provides a good metric realization of the associated Coxeter complex [Ron89, Chap. 2] (in which only spherical facets are represented). Moreover, the metric space under consideration is $CAT(-1)$ if and only if the Coxeter group is Gromov-hyperbolic (which is a weaker property in general [GdlH90, Chap. 3]). Using G. Moussong's non-positively curved complex as a model for apartments, M. Davis [Dav97] proved the existence of a $CAT(0)$-metric realization for any building, which is $CAT(-1)$ whenever the Weyl group of the building is Gromov-hyperbolic. In the latter case, the metric complex is usually not as nice as a hyperbolic tiling as in Proposition 5.1.7.

5.1.5. *Easy non-linearities*

On the one hand, in 5.1.1-5.1.3 we have quoted arguments supporting the analogy between arithmetic groups over function fields and Kac-Moody groups over finite fields. On the other hand, 5.1.4 shows that the geometries are certainly new since hyperbolic buildings can be obtained. Still, it is not proved at this stage that the groups are not well-known groups, only with new actions: what about arguments showing that the groups are new? This question leads to rigidity problems, i.e. proving that groups acting naturally on some geometries cannot act via a big quotient on another geometry. A way to attack the problem is to prove that the groups are not linear over any field. Easy non-linearities are summed up in the following [Rem02a, Theorem 4.6]:

Theorem 5.1.8. *Let Λ be a Kac-Moody group over $\mathbf{F}_q$ with infinite Weyl group, and let Γ be a facet fixator in Λ. Then Γ always contains an infinite group of exponent p. Therefore Γ cannot be linear over any field of characteristic $\neq p$.*

Exhibiting an infinite group of exponent p is made possible thanks to arguments on Kac-Moody root systems and pairs of parallel walls in infinite Coxeter complexes. Then, elementary Zariski closure and algebraic group arguments imply that an infinite group of exponent p cannot be linear over any field of characteristic different from p [Mar91, Lemma VIII.3.7].

5.2. Generalized Kac-Moody groups. Non-isomorphic groups with the same building

In this section, we concentrate on buildings whose apartments are tilings of the hyperbolic plane. Since we are interested in understanding new groups and geometries, it is natural to consider the case of Fuchsian Weyl groups, corresponding to the simplest exotic Kac-Moody groups. The results are mainly of algebraic and combinatorial flavour; they emphasize the role of conditions previously introduced to classify buildings. Abstract isomorphisms between Kac-Moody groups with the same hyperbolic building can be naturally factorized (5.2.1). This is an analogy with the spherical building case, but a difference comes when exhibiting some non-isomorphic groups with the same building (5.2.2). Moreover the local structure of right-angled Fuchsian buildings is particularly simple: this enables one to construct groups which are very close to Kac-Moody groups, but so to speak defined over several ground fields (5.2.3). Back to geometric group theory, this provides lattices of buildings of arbitrarily large rank satisfying a strong non-linearity property (5.2.4).

5.2.1. *Factorization of abstract automorphisms*

We say that a building is *Fuchsian* if its Weyl group is the reflection group of a tiling of the hyperbolic plane (in which case the latter tiling is a pleasant model for apartments). According to Poincaré's Theorem [Mas88, 4.H], Fuchsian tilings are nice metric realizations of Coxeter complexes (the most familiar ones after Euclidean tilings). Elementary facts from hyperbolic geometry enable one to prove the following [Rem02c, Theorem 3.1]:

Theorem 5.2.1. *Let G and G' be two Kac-Moody groups defined over the same finite field $\mathbf{F}_q$ of cardinality $q \geq 4$. Assume that the associated buildings are all isomorphic ($\star$) either to the same locally finite tree, ($\star\star$) or to the same Fuchsian building with regular chambers. Then, up to conjugacy in G, any abstract isomorphism from G to G' is the composition of a permutation of the simple roots and possibly a global opposition of the sign of all roots.*

This factorization result is close to R. Steinberg's classical result on finite groups of Lie type, saying that an abstract automorphism of such a group is the product of a Dynkin diagram automorphism, a ground field automorphism and an inner automorphism [Car72, Theorem 12.5.1]. Stronger factorizations of isomorphisms have recently been obtained by P.-E. Caprace and B. Mühlherr [CM05a], [CM05b].

5.2.2. *Several isomorphism classes*

Let R be a right-angled r-gon of the hyperbolic plane $\mathbb{H}^2$. For any integer $q \geq 2$ there is a unique building $I_{r,q+1}$ whose apartments are Poincaré tilings of $\mathbb{H}^2$ by R, and such that the link at each vertex is the complete bipartite graph of parameters $(q+1, q+1)$ [Bou97, 2.2.1]. Here the *link* of a vertex is a small ball around it, seen as a graph. We call such a building the *right-angled Fuchsian building* of parameters r and $q+1$. These buildings are interesting because they locally look like products of trees, making them simple combinatorially, but globally their Weyl group is irreducible and Fuchsian, leading to remarkable rigidity properties [Bou97], [BP00]. By uniqueness and Proposition 5.1.7, for each $r \geq 5$ the building $I_{r,q+1}$ comes from a Kac-Moody group whenever q is a prime power. Using Theorem 5.2.1, it can be proved that there are abstractly non-isomorphic Kac-Moody groups with the same building [Rem02c, §4, Proposition]:

Corollary 5.2.2. *Let $q \geq 4$ be a prime power and let $r \geq 5$ be an integer. Then there are several abstract isomorphism classes of Kac-Moody groups whose associated buildings are the same $I_{r,q+1}$.*

This result is in contrast with the spherical case, where according to J. Tits' classification, a spherical building of rank ≥ 3 uniquely determines a field and an algebraic group over this field [Tit74]. Here the rank r of the building is an arbitrary integer ≥ 5. This is an argument explaining why in the classification of Moufang twinnings [MR95], [Mue99], the buildings must be assumed to be 2-spherical: Kac-Moody groups with $I_{r,q+1}$ buildings obviously do not satisfy this property.

5.2.3. *Mixing ground fields*

There is an even stronger argument showing that being 2-spherical is a necessary condition for a Moufang twin building to be part of a reasonable classification (where the group side would be given by Kac-Moody groups and their twisted versions). Indeed, some generalizations of Kac-Moody groups with several ground fields can be constructed, provided the buildings have apartments tiled by regular hyperbolic right-angled r-gons. Here «generalization» means that the groups satisfy the same combinatorial axioms as Kac-Moody groups (namely, those of twin root data [Tit92]). We have [RR06, Theorem 3.E]:

Theorem 5.2.3. *A right-angled (Fuchsian) building belongs to a Moufang twinning whenever its thicknesses at panels are cardinalities of projective lines.*

The above quoted reference actually covers more general buildings, namely those buildings whose Weyl group is a right-angled Coxeter group (not necessarily Fuchsian). In the case of trees, a more theoretical construction is sketched by J. Tits in his Notes de Cours au Collège de France [Tit90, §9]. Still, the rank of the infinite dihedral Weyl group there is always equal to 2, whereas in the above result it can be any integer ≥ 5.

5.2.4. *Stronger non-linearities*

Besides its combinatorial interest, the above construction can be seen as a way to produce lattices of hyperbolic buildings with remarkable non-linearity properties. Recall that according to Theorem 5.1.8 the characteristic p of the ground field of a Kac-Moody group prevents this group from being linear over any field of characteristic $\neq p$. Therefore, mixing fields of different characteristics in Theorem 5.2.3 should enable one to produce groups which are not linear over any field. This is indeed the case [RR06, Theorem 4.A]:

Theorem 5.2.4. *Let $r \geq 5$ be an integer and $\{\mathbf{K}_i\}_{i \in \mathbf{Z}/r}$ be a family of fields, among which are two fields with different positive characteristics. Let Λ be a group defined as in Theorem 5.2.3 from these fields and let Γ be a chamber fixator. Then, any group homomorphism $\rho : \Gamma \to \prod_{\alpha \in A} \mathbf{G}_\alpha(\mathbf{F}_\alpha)$ has infinite kernel, whenever the index set A is finite and $\mathbf{G}_\alpha$ is a linear algebraic group defined over the field $\mathbf{F}_\alpha$ for each $\alpha \in A$.*

Note that the result is stronger than plain non-linearity since mixing ground fields is also allowed at the (linear) right hand-side of the representation. Moreover the kernel is not only non-trivial, but always infinite. From the purely Kac-Moody viewpoint, this result says that the truly difficult case of non-linearity is when the characteristic of the target algebraic group is the characteristic of the ground field of the Kac-Moody group. This case is reviewed in Sect. 5.4.

5.3. Generalized algebraic groups over local fields

A Kac-Moody group Λ acts discretely on the product of its buildings, but its action on a single factor is no longer discrete. Therefore it makes sense to take the closure $\bar{\Lambda} < \mathrm{Aut}(X_\pm)$ of the image of a Kac-Moody group in such a non-discrete action. The result is called a *topological Kac-Moody group* (the kernel of the Λ-action on $X_\pm$ is the finite center of Λ). In the classical case $\Lambda = \mathrm{SL}_n(\mathbf{F}_q[t, t^{-1}])$, $X_\pm$ is the Bruhat-Tits building of $\mathrm{SL}_n(\mathbf{F}_q((t)))$ or $\mathrm{SL}_n(\mathbf{F}_q((t^{-1})))$, respectively. If $\mu_n(\mathbf{F}_q)$ denotes the n-th roots of unity in $\mathbf{F}_q$, the image $\Lambda/Z(\Lambda)$ of $\mathrm{SL}_n(\mathbf{F}_q[t, t^{-1}])$ under the action on $X_\pm$ is $\mathrm{SL}_n(\mathbf{F}_q[t, t^{-1}])/\mu_n(\mathbf{F}_q)$ and the completions $\bar{\Lambda}$ are $\mathrm{PSL}_n(\mathbf{F}_q((t)))$ and $\mathrm{PSL}_n(\mathbf{F}_q((t^{-1})))$, respectively. In fact, there are many arguments to compare a topological Kac-Moody group with a semisimple group over a local field: existence of a combinatorial structure refining Tits systems (5.3.1), group-theoretic characterization of chamber-fixators in terms of pro-p subgroups (5.3.2), topological simplicity

(5.3.3). Still, as for discrete groups we have phenomena suggesting that some so-obtained totally disconnected groups are new (5.3.4). We finally quote a result exhibiting the coexistence of non-linear non-uniform lattices and uniform lattices embedding convex-cocompactly in hyperbolic spaces (5.3.5).

5.3.1. *Refined Tits systems*

The structure of a *refined Tits system* is due to V. Kac and D. Peterson [KP85]; it is a generalization of a split BN-pair, a notion from the theory of finite groups of Lie type. The difference with a plain Tits system is the formalization in the axioms of the existence of an abstract unipotent subgroup in the Borel subgroup. We have [RR06, Theorem 1.C]:

Theorem 5.3.1. *Let Λ be a Kac-Moody group over the finite field $\mathbf{F}_q$ of characteristic p. The associated topological Kac-Moody group $\overline{\Lambda}$ admits a refined Tits system, and the Tits system gives rise to a building in which any facet-fixator is, up to finite index, a pro-p group.*

It is well-known that a group acting strongly transitively on a building admits a natural Tits system [Ron89, Theorem 5.2], so the point in the first assertion lies in the difference between a Tits system and a refined Tits system. Moreover standard properties of double cosets in Bruhat decompositions imply that topological Kac-Moody group are compactly generated [Rem04b, Corollary 1.B.1]. It is explained in [Rem04b, 1.B.1] why refined Tits systems for $\overline{\Lambda}$ imply the existence of a lot of torsion in facet fixators – called *parahoric subgroups*. This explains why the analogy with semisimple groups over local fields is relevant only when the local field has the same characteristic p as the finite ground field of Λ (parahoric subgroups of semisimple groups over finite extensions of $\mathbf{Q}_l$ are virtually torsion free). In analogy with the classical case of Bruhat-Tits theory [BT72], [BT84], we call *Iwahori subgroup* a chamber fixator in $\overline{\Lambda}$.

5.3.2. *Iwahori subgroups*

Back to the case $\Lambda = \mathrm{SL}_n(\mathbf{F}_q[t, t^{-1}])$, let v be a vertex in the Bruhat-Tits building of $\mathrm{SL}_n(\mathbf{F}_q((t)))$. Then its fixator is isomorphic to $\mathrm{SL}_n(\mathbf{F}_q[[t]])$, and for some chamber containing v the corresponding Iwahori subgroup is the group of matrices in $\mathrm{SL}_n(\mathbf{F}_q[[t]])$ reducing to upper triangular matrices modulo t. Moreover the first congruence subgroup of $\mathrm{SL}_n(\mathbf{F}_q[[t]])$, i.e. the matrices reducing to the identity modulo t, is a maximal pro-p subgroup whose normalizer is the above Iwahori subgroup. In the general Kac-Moody setting, the result below [Rem04b, Proposition 1.B.2] is a generalization of [PR94, Theorem 3.10].

Proposition 5.3.2. *The Iwahori subgroups are group-theoretically characterized as the normalizers of the pro-p Sylow subgroups.*

In general, the analogue of the first congruence subgroup of a vertex fixator must be defined as the (pointwise) fixator of the star around the vertex under consideration.

5.3.3. *Topological simplicity*

In view of the simplicity of adjoint algebraic simple groups over large enough fields, the theorem below [Rem04b, Theorem 2.A.1] is not surprising:

Theorem 5.3.3. *For $q \geq 4$, a topological Kac-Moody group over a finite field is the direct product of finitely many topologically simple groups, with one factor for each connected component of its Dynkin diagram.*

This result is extremely useful when extending abstract representations of finitely generated Kac-Moody groups to continuous representations of topological Kac-Moody groups (Theorem 5.4.3). The arguments of the proof are basically: a normal subgroup in an irreducible Tits system either is chamber-transitive or acts trivially on the associated building [Bou81, IV.2], Iwahori subgroups are virtually pro-p, and a generating set for a Kac-Moody group can be chosen in a finite collection of finite subgroups of Lie type (which are perfect whenever $q \geq 4$). It is reasonable to think that topological Kac-Moody groups are in fact abstractly simple (Question 5.5.7).

5.3.4. *Non-homogeneous Furstenberg boundaries*

When the building $X_\pm$ has hyperbolic apartments, the existence of many hyperbolic translations leads to an interesting connection with topological dynamics. Let Y be a compact metrizable space and let us denote by $\mathscr{M}^1(Y)$ the space of probability measures on Y. Recall that $\mathscr{M}^1(Y)$ is compact and metrizable for the weak-$*$ topology. If Y admits a continuous action by a topological group G, we say that Y is a *Furstenberg boundary* for G if it is G*-minimal* and G*-strongly proximal* [Mar91, VI.1.5]. The first condition says that any G-orbit in Y is dense and the second one says that any G-orbit closure in $\mathscr{M}^1(Y)$ contains a Dirac mass. The arguments of [Rem04b, Lemma 4.B.1] give:

Lemma 5.3.4. *Let Λ be a Kac-Moody group whose buildings have apartments isomorphic to a hyperbolic tiling. Then the asymptotic boundary $\partial_\infty X_\pm$ is a Furstenberg boundary for any closed automorphism group of $X_\pm$ containing $\bar{\Lambda}$.*

This existence of non-homogeneous boundaries is new with respect to semisimple Lie groups (archimedean or not), since in the latter case any Furstenberg boundary is equivariantly isomorphic to a flag variety of the group [GJT98, 9.37].

5.3.5. *Coexistence of two kinds of lattices*

We close this section by stating a result which says that some topological Kac-Moody groups contain lattices of surprisingly different nature [RR06, Proposition 4.B].

Proposition 5.3.5. *There exist topological Kac-Moody groups $\bar{\Lambda}$ over $\mathbf{F}_q$ which contain both non-uniform lattices which cannot be linear over any field of characteristic prime to q, and uniform lattices which have convex-cocompact embeddings into real hyperbolic spaces. The limit sets of the latter embeddings often have Hausdorff dimension > 2.*

It follows from [BM96, Corollary 0.5] that if a Kac-Moody group Λ is S-arithmetic and such that $\bar{\Lambda}$ is a higher-rank simple Lie group, then any lattice of $\bar{\Lambda}$ fixes a point in each of its actions on proper $CAT(-1)$-spaces. Therefore the above phenomenon is excluded in the classical algebraic situation, unless the building of $\bar{\Lambda}$ is a tree, meaning that the Weyl group Λ is infinite dihedral (of rank 2 as a Coxeter group). The rank r of a Fuchsian building $I_{r,q+1}$ may be chosen arbitrarily large ≥ 5.

5.4. Non-linearity in equal characteristics

In view of the easy non-linearity results (Theorem 5.1.8), the remaining linearity to disprove is for Kac-Moody groups over finite fields of characteristic p into linear groups of characteristic p. But some Kac-Moody S-arithmetic groups such as $\mathrm{SL}_n(\mathbf{F}_q[t,t^{-1}])$ are linear in equal characteristic. Therefore the question is rather, for any finite field of characteristic p, to find examples of Kac-Moody groups that cannot be linear over any field of characteristic p either. The main idea is to use some superrigidity property (5.4.1) to show that the existence of a faithful abstract homomorphism from a finitely generated Kac-Moody group into an algebraic group implies the existence of an embedding of a topological Kac-Moody group into a non-Archimedean simple group (5.4.2). The existence of such an embedding is expected to be simpler to disprove because it comes with an embedding of the vertices of the (possibly exotic) Kac-Moody building to a Euclidean one, which enables one to take advantage of the incompatibility between hyperbolic and Euclidean geometries. For some Kac-Moody groups with Fuchsian buildings this is indeed the case (5.4.3). The same circle of ideas allows us to show a complementary result: a non-faithful representation from a finitely generated Kac-Moody group most of the time has a virtually solvable image (5.4.4). Combining the latter results leads to groups all of whose linear images are finite (5.4.5).

5.4.1. *Commensurator superrigidity*

Let us first recall that the *commensurator* of a group inclusion $\Delta < G$ is the group:

$$\mathrm{Comm}_G(\Delta) = \{g \in G \mid \Delta \cap g\Delta g^{-1} \text{ has finite index in both } \Delta \text{ and } g\Delta g^{-1}\}.$$

The next theorem is basically due to G.A. Margulis [Mar91, VII.5.4], and the formulation below can be found in [Bon, Theorem 1]. G.A. Margulis proved it when G is a semisimple group over a local field. A (certainly non-exhaustive) list of later contributions is the following: in [AB94], the semisimple group G is replaced by a group containing an amenable subgroup $P < G$ similar to a minimal parabolic subgroup; in [Bur95] the existence of P is replaced by the existence of a substitute for a Furstenberg boundary; in [BM02] such boundaries are constructed for compactly generated groups G; and the double ergodicity Theorem in [Kai02] shows that suitable boundaries are available for any locally compact second countable group, via Poisson boundary theory. Different approaches lead to similar results: [Mar94] by means of equivariant generalized harmonic mappings and [Sha00] by means of representation theory.

Theorem 5.4.1. *Let G be a locally compact second countable topological group, $\Gamma < G$ be a lattice and Λ be a subgroup of G with $\Gamma < \Lambda < \mathrm{Comm}_G(\Gamma)$. Let k be a local field and H be a connected almost k-simple algebraic group. Assume that $\pi : \Lambda \to H_k$ is a homomorphism such that $\pi(\Lambda)$ is dense in the Zariski topology on H and $\pi(\Gamma)$ is unbounded in the Hausdorff topology on H_k. Then π extends to a continuous homomorphism $\bar{\Lambda} \to H_k/Z(H_k)$, where $Z(H_k)$ is the center of H_k.*

A more difficult superrigidity result consists in extending representations of lattices (instead of their commensurators) to continuous representations of the ambient topological group (to algebraic representations when so is the ambient group). This is a more difficult result which requires stronger assumptions (e.g. higher rank for algebraic ambient groups), and again the main ideas are due to G.A. Margulis. The first results in positive characteristic were proved by T.N. Venkataramana [Ven88].

5.4.2. *Embedding theorem*

For the following theorem, we were inspired by a paper due to A. Lubotzky, Sh. Mozes and R.J. Zimmer [LMZ94], where superrigidity is used to disprove linearities for commensurators of tree lattices. Trees are one-dimensional buildings, and Kac-Moody theory naturally leads to lattices for buildings of arbitrary dimension. We have [RR06, 1.B Lemma 2]:

Lemma 5.4.2. *Let Λ be any Kac-Moody group over $\mathbf{F}_q$ and let Γ be any negative facet fixator in Λ. Then: $\Lambda < \mathrm{Comm}_{\bar{\Lambda}}(\Gamma)$.*

Therefore it makes sense to use commensurator superrigidity because when $q \gg 1$, Γ is a lattice of $\bar{\Lambda}$ by Theorem 5.1.3. This enables to obtain [Rem04b, §3, Theorem]:

Theorem 5.4.3. *Let Λ be a Kac-Moody group over the finite field $\mathbf{F}_q$ of characteristic p with $q > 4$ elements, with infinite Weyl group W and buildings X_+ and X_-. Let $\bar{\Lambda}$ be the corresponding Kac-Moody topological group. We make the following assumptions:*

(TS) *the group $\bar{\Lambda}$ is topologically simple;*
(NA) *the group $\bar{\Lambda}$ is not amenable;*
(LT) *the group Λ is a lattice of $X_+ \times X_-$ for its diagonal action.*

Then, if Λ is linear over a field of characteristic p, there exists:

- *a local field k of characteristic p and a connected adjoint k-simple group $\mathbf{G}$,*
- *a topological embedding $\mu : \bar{\Lambda} \to \mathbf{G}(k)$ with Hausdorff unbounded and Zariski dense image,*
- *and a μ-equivariant embedding $\iota : V_{X_+} \to V_\Delta$ from the set of vertices of the Kac-Moody building X_+ of Λ to the set of vertices of the Bruhat-Tits building Δ of $\mathbf{G}(k)$.*

Conditions (TS) and (LT) are satisfied whenever the Weyl group W of Λ is irreducible and $q \gg 1$. Note that the conclusion of Theorem 5.4.3 provides a continuous extension but it is not clear that this topological homomorphism is a closed map.

Indeed, there is still work to do [Rem04b, Lemma 3.C] and, besides the topological simplicity of $\bar{\Lambda}$, the arguments used for that are of combinatorial nature: basically the Bruhat decomposition of $\bar{\Lambda}$ with respect to an Iwahori subgroup.

5.4.3. *Non-linear Fuchsian Kac-Moody groups*

In the case of some Kac-Moody groups with Fuchsian buildings, it can be proved that the associated topological groups cannot be closed subgroups of any simple non-Archimedean Lie group, thus implying the non-linearity of the involved finitely generated groups [Rem04b, Theorem 4.C.1].

Theorem 5.4.4. *Let Λ be a countable Kac-Moody group over $\mathbf{F}_q$ with right-angled Fuchsian associated buildings. Assume that any prenilpotent pair of roots not contained in a spherical root system leads to a trivial commutation of the corresponding root groups. Then for $q \gg 1$, the group Λ is not linear over any field.*

The condition on commutation of root groups is technical: prenilpotency of pairs of roots is relevant to abstract root systems of Coxeter groups [Tit87], [Rem02b, 1.4.1]. It is not very restrictive and actually a weaker assumption may be required – see the remark after [Rem04b, Lemma 4.A.2]. Seeing roots as half-apartments, a pair of roots is prenilpotent if and only if the walls of the roots intersect or if a root contains the other.

The main idea at this stage is to use a dynamical characterization of parabolic subgroups proved by G. Prasad in a paper on strong approximation [Pra77, Lemma 2.4] in the algebraic case, e.g. the case of the target $\mathbf{G}(k)$ of the continuous extension $\mu : \bar{\Lambda} \to \mathbf{G}(k)$. In the case of hyperbolic buildings, one must first say that a parabolic subgroup is by definition a boundary point fixator (in analogy with the symmetric space or Bruhat-Tits building case), and then show that the dynamical characterization holds too [Rem04b, Lemma 4.B.2]. We now have a dynamical round about to the non-existence of an algebraic structure on $\bar{\Lambda}$, and this enables one to show that under the continuous homomorphism μ, parabolic subgroups go to parabolic subgroups. The contradiction comes when one notes that, thanks to the strong dynamics of a Fuchsian group on the asymptotic boundary of the hyperbolic plane, the dynamical analogues of unipotent radicals on the left hand-side are not normalized by the parabolic subgroups containing them, whereas it is the case by definition on the right hand-side of μ.

5.4.4. *Virtual solvability of non-faithful linear images*

In 5.4.2, the starting point is a faithful abstract representation from a finitely generated Kac-Moody group. Starting from non-faithful representations is interesting too, since it can be shown that in this case the images are solvable up to finite index. In other words, when the kernel is non-trivial, it is big [Rem05, Theorem 11]:

Theorem 5.4.5. *Let Λ be a Kac-Moody group over the finite field $\mathbf{F}_q$ of characteristic p, with connected Dynkin diagram and $q \gg 1$. Let $\rho : \Lambda \to \mathrm{GL}_n(\mathbf{F})$ be a linear representation. If ρ is not faithful, the group $\rho(\Lambda)$ is virtually solvable. In particular if Λ is Kazhdan, $\rho(\Lambda)$ is finite.*

Recall that by Theorem 5.1.6, many Kac-Moody groups over $\mathbf{F}_q$ are Kazhdan whenever $q \gg 1$. The above theorem is proved thanks to the same kind of arguments as for Theorem 5.4.3. In order to be in a position to apply Theorem 5.4.1, we have to take the Zariski closure of the image $\rho(\Lambda)$ and to mod out by the radical $R(\overline{\rho(\Lambda)}^Z)$ of the latter algebraic group. It then suffices to show that the image of Λ in $\overline{\rho(\Lambda)}^Z / R(\overline{\rho(\Lambda)}^Z)$ is finite, which can be done thanks to Burnside's theorem [Jac89, 4.5, Exercise 8]. A way to fulfill the unboundedness assumption on the image of Λ (in the Hausdorff topology) is to use tricks from Tits's paper on the existence of free groups in linear groups [Tit72].

5.4.5. *Groups all of whose linear images are finite*

From the previous result, it can be shown that some Kac-Moody groups with hyperbolic buildings don't have any infinite linear image [Rem05, Theorem 16].

Theorem 5.4.6. *There is an integer N such that for any Kac-Moody group Λ over $\mathbf{F}_q$ whose buildings have apartments isomorphic to the tessellation of the hyperbolic plane by regular triangles of angle $\frac{\pi}{4}$, if $q \geq N$ then any linear image of Λ is finite, whatever the target field.*

The proof roughly goes as follows. By Theorem 5.4.5, it is enough to show that some Kac-Moody groups enjoying Kazhdan's property (T) are not linear over any field. According to Theorem 5.1.6, the groups as in the theorem are Kazhdan, and by twin root datum arguments it can be proved that such groups contain Kac-Moody-like subgroups to which the proof of Theorem 5.4.4 applies.

5.5. Conjectures and questions

In this final section, we propose a few questions about finitely generated and totally disconnected Kac-Moody groups. Here, implicit in the word *totally disconnected* is that the group under consideration is uncountable and not endowed with the discrete topology.

5.5.1. *Finitely generated groups*

For discrete groups, we think that the class of non-linear Kac-Moody groups is much wider than the one for which the property has been proved so far.

Conjecture 5.5.1. If the Weyl group W is Gromov-hyperbolic and if $q \gg 1$, then the Kac-Moody group Λ is not linear.

Recall that according to G. Moussong, in the class of Coxeter groups hyperbolicity is equivalent to acting discretely and cocompactly on a $CAT(-1)$-space [Mou88]. The non-linearity proof of Sect. 5.4 deals with groups Λ whose Weyl groups have pleasant Coxeter complexes since the latter complexes are Fuchsian tilings. In general,

Moussong's complex is defined by isometric gluings of cells and is less easy to understand, so one may have to use arguments of combinatorial nature on Kac-Moody root systems at some points. Note also that one should get rid of the technical condition on prenilpotent pairs of roots in the statement of Theorem 5.4.4.

Another natural question is:

Question 5.5.2. Can the assumption «$q \gg 1$» be removed in non-linearity results?

In other words: is there a generalized Cartan matrix A and a prime number p such that the corresponding group is linear over $\mathbf{Z}/p$ but no longer linear over $\mathbf{F}_{p^r}$ for large enough r? Note that for Λ to be a lattice of the product of its buildings – which is a crucial argument in our proof - the value of q is important (Theorem 5.1.3). So if the answer to the above problem is yes, the non-linearity proof should require new ideas. Moreover a counter-example due to P. Abramenko shows that for buildings whose chambers are regular hyperbolic triangles of angles $\frac{\pi}{4}$, facet fixators in some groups over $\mathbf{F}_2$ or $\mathbf{F}_3$ are not finitely generated [Abr03, Counter-example 1, Remark 2], hence cannot have property (T) [Mar91, Theorem III.2.7]. But according to Theorem 5.1.6, for $q \gg 1$ the groups do enjoy Kazhdan's property (T).

The most general question on non-linearity for Kac-Moody groups is:

Problem 5.5.3. Find necessary and sufficient conditions for the non-linearity of a finitely generated Kac-Moody group, only in terms of the generalized Cartan matrix defining the group.

Of course, if the generalized Cartan matrix A is of affine type, meaning that the Weyl group is a Euclidean reflection group, then the corresponding Kac-Moody groups are S-arithmetic groups, hence are linear. A restatement is: are there other linear examples than the affine ones?

Finally, there is another problem which comes from the analogy with the situation of lattices, namely the problem of arithmeticity. Of course when algebraic structures are available, the definition is well-known [Zim84, Definitions 6.1.1 and 10.1.11]. In the general context of an inclusion $\Delta < G$ of a discrete subgroup Δ in a locally compact group G, it has now become classical to say that Δ is *arithmetic* in G if by definition $\mathrm{Comm}_G(\Delta)$ is dense in G (5.4.1). This is reasonable since by Margulis's commensurator criterion, a lattice in a non-compact simple Lie group is arithmetic in the classical sense if and only if it is so in the above sense [Zim84, Theorem 6.2.5]. This definition makes sense in the case where the topological group G is the isometry group of some metric space, e.g. of some building with enough automorphisms. Since Kac-Moody buildings are close in many respects to Bruhat-Tits buildings, it is natural to ask:

Question 5.5.4. Let X_+ be the positive building of some Kac-Moody group Λ over $\mathbf{F}_q$. Is a negative facet fixator in Λ arithmetic in $\mathrm{Aut}(X_+)$? More generally, which lattices, not necessarily relevant to Kac-Moody groups, are arithmetic in $\mathrm{Aut}(X_+)$?

Note that by definition of a topological Kac-Moody group (Sect. 5.3, introduction), any negative facet fixator Γ is arithmetic in $\bar{\Lambda}$ since by Lemma 5.4.2, $\mathrm{Comm}_{\bar{\Lambda}}(\Gamma)$ contains Λ. Here is a list of known arithmeticity results:

(i) Cocompact lattices in a locally finite tree T are all arithmetic in Aut(T) [Liu94].
(ii) The Nagao lattice $SL_2(\mathbf{F}_q[t^{-1}])$ is a non-uniform arithmetic lattice in the full automorphism group of the Bruhat-Tits tree of $SL_2(\mathbf{F}_q((t)))$ [Moz99]. This was later extended to some Moufang twin trees, which enables to deal with non-linear non-uniform tree lattices [AR06].
(iii) Uniform lattices in some hyperbolic buildings, not necessarily from Kac-Moody theory [Hag03].

See the end of [AR06] for more precise questions.

We finish by another question about linearity, which deals with facet fixators in finitely generated Kac-Moody groups instead of the full latter groups.

Question 5.5.5. Let Λ be a Kac-Moody group over $\mathbf{F}_q$. Are there non-linear facet fixators in Λ?

This question seems to be more difficult because the latter groups are always residually finite (linearity of finitely generated groups is often disproved by disproving residual finiteness, i.e. using the Mal'cev theorem). To see the residual finiteness in this case, it suffices to take subgroups fixing combinatorial balls of increasing radius around the given facet; the intersection of these groups is trivial since it fixes pointwise the building. Constructing and studying non-linear finitely generated groups acting on $CAT(-1)$ metric spaces is a very interesting problem. Note that the above facet fixators are often, but not always, finitely generated.

5.5.2. *Totally disconnected groups*

Theorem 5.4.3 shows that non-linearity of some finitely generated groups roughly amounts to non-linearity of much bigger topological groups. In the study of topological Kac-Moody groups in Sect. 5.3 appear an interesting class of pro-p subgroups. A way to disprove some linearities would be to answer the following:

Question 5.5.6. Which topological Kac-Moody groups contain non-linear pro-p subgroups?

Forgetting discrete groups, the question is interesting in its own right: for instance, the linearity of free pro-p subgroups is a question so far admitting only partial answers (E. Zelmanov).

In the classical case, topological Kac-Moody groups correspond to groups of rational points of adjoint semisimple groups, so it makes sense to ask the following:

Question 5.5.7. Are topological Kac-Moody groups direct products of abstractly simple groups?

Theorem 5.3.3 only says that the latter groups are direct products of topologically simple groups. F. Haglund and F. Paulin, elaborating on J. Tits' proof for trees [Tit70], showed the abstract simplicity of full automorphism groups of many hyperbolic buildings [HP98, Theorems 1.1 and 1.2].

References

[AB94] N. A'Campo and M. Burger, Réseaux arithmétiques et commensurateur d'après G.A. Margulis, *Inventiones Mathematicae*, **116** (1994) 1–25.

[Abr97] P. Abramenko, Twin buildings and applications to S-arithmetic groups, Lecture Notes in Mathematics, vol. 1641, Springer (1997).

[Abr03] P. Abramenko, Finiteness properties for groups acting on twin buildings, Groups: Topological, Combinatorial and Arithmetic Aspects (Bielefeld, 1999) (T. W. Müller, ed.), LMS Lecture Notes Series, London Mathematical Society, Cambridge University Press (2003) pp. 17–21.

[AM97] P. Abramenko and B. Mühlherr, Présentations de certaines BN-paires jumelées comme sommes amalgamées, *Comptes-Rendus de l'Académie des sciences de Paris*, **325** (1997) 701–706.

[AR06] P. Abramenko and B. Remy, Commensurators of some nonuniform tree lattices and Moufang twin trees, this volume (2006).

[Bar00] S. Barré, Immeubles de Tits triangulaires exotiques, *Ann. Fac. Sci. Toulouse Math.*, **9** (2000) no. 4, 575–603.

[Beh69] H. Behr, Endliche Erzeugbarkeit arithmetischer Gruppen über Funktionenkörpern, *Inventiones Mathematicæ*, **7** (1969) 1–32.

[Beh03] H. Behr, Higher finiteness properties for S-arithmetic groups in the function field case I, Groups: Topological, Combinatorial and Arithmetic Aspects (Bielefeld, 1999) (T. W. Müller, ed.), LMS Lecture Notes Series, London Mathematical Society, Cambridge University Press (2003) pp. 761–769.

[BH99] M. Bridson and A. Haefliger, Metric spaces of non-positive curvature, Grundleheren der mathematischen Wissenschaften, vol. 319, Springer (1999).

[BM96] M. Burger and Sh. Mozes, CAT(-1)-spaces, divergence groups and their commensurators, *Journal of the American Mathematical Society*, **9** (1996) 57–93.

[BM97] M. Burger and Sh. Mozes, Finitely presented simple groups and products of trees, *Comptes-Rendus de l'Académie des sciences de Paris*, **324** (1997) 747–752.

[BM00a] M. Burger and Sh. Mozes, Groups acting on trees: from local to global structure, *Publications Mathématiques de l'IHÉS*, **92** (2000) 113–150.

[BM00b] M. Burger and Sh. Mozes, Lattices in product of trees, *Publications Mathématiques de l'IHÉS*, **92** (2000) 151–194.

[BM02] M. Burger and N. Monod, Continuous bounded cohomology and applications to rigidity theory, *Geometric and Functional Analysis*, **12** (2002) no. 2, 219–280.

[Bon] P. Bonvin, Strong boundaries and commensurator super-rigidity, appendix to [Rem04b].

[Bou81] N. Bourbaki, Groupes et algèbres de Lie IV–VI, Éléments de mathématique, Masson (1981).

[Bou97] M. Bourdon, Immeubles hyperboliques, dimension conforme et rigidité de Mostow, *Geometric and Functional Analysis*, **7** (1997) 245–268.

[BP00] M. Bourdon and Hervé Pajot, Rigidity of quasi-isometries for some hyperbolic buildings, *Commentarii Mathematici Helvetici*, **75** (2000) 701–736.

[Bro87] K. S. Brown, Cohomology of groups, Graduate texts in mathematics, vol. 82, Springer (1987).

[Bro89] K. S. Brown, Buildings, Springer (1989).

[BT72] F. Bruhat and J. Tits, Groupes réductifs sur un corps local I. Données radicielles valuées, *Publications Mathématiques de l'IHÉS*, **41** (1972) 5–251.

[BT84] F. Bruhat and J. Tits, Groupes réductifs sur un corps local II. Schémas en groupes. Existence d'une donnée radicielle valuée, *Publications Mathématiques de l'IHÉS*, **60** (1984) 5–184.

[Bur95] M. Burger, Rigidity properties of group actions on CAT(0)-spaces, Proceedings of the International Congress of Mathematicians (Zürich, August 1994) (S. D. Chatterji, ed.), Birkhäuser Verlag, Basel (1995) pp. 761–769.

[Car72] R. W. Carter, Simple groups of Lie type, Wiley Intersciences (1972).

[CG03] L. Carbone and H. Garland, Existence of lattices in Kac-Moody groups over finite fields, *Commun. Contemp. Math.*, **5** (2003) 813–867.

[CM05a] P.-E. Caprace and B. Muehlherr, Isomorphisms of Kac-Moody groups, *Invent. Math.*, **161** (2005) 361–388.

[CM05b] P.-E. Caprace and B. Muehlherr, Isomorphisms of Kac-Moody groups which preserve bounded subgroups, Adv. Math., to appear (2005).

[Dav97] M. Davis, Buildings are CAT(0), Geometry and cohomology in group theory (P. H. Kropholler, G. A. Niblo, and R. Stöhr, eds.), LMS Lecture Notes Series 252, London Mathematical Society, Cambridge University Press, 1997, pp. 108–123.

[DJ02] J. Dymara and T. Januszkiewicz, Equivariant cohomology of buildings and of their automorphism groups, *Inventiones Mathematicæ*, **150** (2002) no. 3, 579–627.

[dlHV89] P. de la Harpe and A. Valette, La propriété (T) de Kazhdan pour les groupes localement compacts, Astérisque, vol. 175, Société Mathématique de France (1989).

[GdlH90] É. Ghys and P. de la Harpe, Sur les groupes hyperboliques d'après Mikhael Gromov, Progress in Mathematics, vol. 83, Birkhäuser (1990).

[GJT98] Y. Guivarc'h, L. Ji, and J. C. Taylor, Compactifications of symmetric spaces, Progress in Mathematics, vol. 156, Birkhäuser (1998).

[Hag03] F. Haglund, Linearity and commensurability of uniform lattices of right-angled hyperbolic buildings, private communication (2003).

[Har69] G. Harder, Minkowskische Reduktionstheorie über Funktionenkörpern, *Inventiones Mathematicæ*, **7** (1969) 33–54.

[HP98] F. Haglund and F. Paulin, Simplicité de groupes d'automorphismes d'espaces à courbure négative, *Geom. Topol. Monogr.*, **1** (1998) 181–248.

[Jac89] N. Jacobson, Basic Algebra II (2nd edition), Freeman and Co. (1989).

[Kai02] V. A. Kaimanovich, Double ergodicity of the Poisson boundary and applications to bounded cohomology, to appear in Geometric and Functional Analysis (2002).

[KP85] V. Kac and D. Peterson, Defining relations for certain infinite-dimensional groups, Élie Cartan et les mathématiques d'aujourd'hui. The mathematical heritage of Élie Cartan. Lyon, 25–29 juin 1984 (Société mathématique de France, ed.), Astérisque Hors-Série (1985) pp. 165–208.

[Liu94] Y. Liu, Density of the commensurability group of uniform tree lattices, *Journal of Algebra*, **165** (1994) 346–359.

[LMZ94] A. Lubotzky, Sh. Mozes and R. J. Zimmer, Superrigidity for the commensurability group of tree lattices, *Commentarii Mathematici Helvetici*, **69** (1994) 523–548.

[Mar91] G. A. Margulis, Discrete Subgroups of Semisimple Lie Groups, Ergebnisse der Mathematik und ihrer Grenzgebiete (3), vol. 17, Springer (1991).

[Mar94] G. A. Margulis, Superrigidity of commensurability subgroups and generalized harmonic maps, unpublished (1994).

[Mas88] B. Maskit, Kleinian groups, Grundlehren der matematischen Wissenschaften, vol. 287, Springer (1988).

[Mou88] G. Moussong, Hyperbolic Coxeter groups, PhD thesis, Ohio State University (1988).

[Moz99] Sh. Mozes, Trees, lattices and commensurators, Algebra, K-Theory, Groups, and Education: On the Occasion of Hyman Bass's 65th Birthday (T. Y. Lam and A. R. Magid, eds.), Contemporary Mathematics, no. 243, American Mathematical Society (1999) pp. 145–151.

[MR95] B. Mühlherr and M. A. Ronan, Local to global structure in twin buildings, *Inventiones Mathematicæ*, **122** (1995) 71–81.

[Mue99] B. Muehlherr, Locally split and locally finite buildings of 2-spherical type, *J. Reine angew. Math.*, **511** (1999) 119–143.

[PR94] V. Platonov and A. Rapinchuk, Algebraic groups and number theory, Academic press (1994).

[Pra77] G. Prasad, Strong approximation for semi-simple groups over function fields, *Annals of Mathematics*, **105** (1977) 553–572.

[Rem99] B. Remy, Construction de réseaux en théorie de Kac-Moody, *Comptes-Rendus de l'Académie des sciences de Paris*, **329** (1999) 475–478.

[Rem02a] B. Remy, Classical and non-linearity properties of Kac-Moody groups, Rigidity in Dynamics and Geometry (Newton Institute, 2000) (M. Burger and A. Iozzi, eds.), Springer Verlag (2002) pp. 391–406.

[Rem02b] B. Remy, Groupes de Kac-Moody déployés et presque déployés, Astérisque, vol. 277, Société Mathématique de France (2002).

[Rem02c] B. Remy, Immeubles de Kac-Moody hyperboliques. Groupes non isomorphes de même immeuble, *Geometriae Dedicata*, **90** (2002) 29–44.

[Rem04a] B. Remy, Kac-Moody groups: split and relative theories. Lattices, Groups: Topological, Combinatorial and Arithmetic Aspects (Bielefeld, 1999) (T. W. Müller, ed.), LMS Lecture Notes Series, London Mathematical Society, Cambridge University Press (2004) pp. 487–541.

[Rem04b] B. Remy, Topological simplicity, commensurator super-rigidity and non-linearities of Kac-Moody groups, *Geometric and Functional Analysis*, **14** (2004) 810–852.

[Rem05] B. Remy, Linear images of Kac-Moody groups, in preparation (2005).

[Ron86] M. A. Ronan, A construction of buildings with no rank 3 residues of spherical type, Buildings and the geometry of diagrams (Lectures given at the third 1984 session of the Centro Internazionale Matematico Estivo (CIME) held in Como, August 26–September 4, 1984) (L. A. Rosati, ed.), Lecture Notes in Mathematics, vol. 1181, Springer Verlag, Berlin (1986) pp. 242–248.

[Ron89] M. A. Ronan, Lectures on Buildings, Perspectives in Mathematics, vol. 7, Academic Press (1989).

[RR06] B. Remy and M. Ronan, Topological groups of Kac-Moody type, right-angled twinnings and their lattices, *Commentarii Math. Helv.*, **81** (2006) 191–219.

[Sha00] Y. Shalom, Rigidity of commensurators and irreducible lattices, *Inventiones Mathematicæ*, **141** (2000) 1–54.

[Tit70] J. Tits, Sur le groupe des automorphismes d'un arbre, Essays on topology and related topics (Mémoires dédiés à Georges de Rham (A. Hæfliger and R. Narasimhan, eds.), Springer Verlag (1970) pp. 188–211.

[Tit72] J. Tits, Free subgroups in linear groups, *Journal of Algebra*, **20** (1972) 250–270.

[Tit74] J. Tits, Buildings of spherical type and finite BN-pairs, Lecture Notes in Mathematics, vol. 386, Springer (1974).

[Tit87] J. Tits, Uniqueness and presentation of Kac-Moody groups over fields, *Journal of Algebra*, **105** (1987) 542–573.

[Tit89] J. Tits, Groupes associés aux algèbres de Kac-Moody, Séminaire Bourbaki 700, Astérisque, vol. 177–178, Société Mathématique de France (1989) pp. 7–31.

[Tit90] J. Tits, Théorie des groupes, Annuaire Collège de France (Résumé de cours) (1990) pp. 87–103.

[Tit92] J. Tits, Twin buildings and groups of Kac-Moody type, Groups, Combinatorics & Geometry (LMS Symposium on Groups and Combinatorics, Durham, July 1990) (M. Liebeck and J. Saxl, eds.), LMS Lecture Notes Series, vol. 165, Cambridge University Press (1992) pp. 249–286.

[Ven88] T. N. Venkataramana, On superrigidity and arithmeticity of lattices in semisimple groups over local fields of arbitrary characteristic, *Inventiones Mathematicæ*, **92** (1988) no. 2, 255–302.

[Zim84] R. J. Zimmer, Ergodic Theory and Semisimple Groups, Monographs in Mathematics, vol. 81, Birkhäuser (1984).

Proc. Inst. Wrk.shps. Int. Conf. – Essays in Geometric Group Theory No. 9, 2009, pp. 125–143.

Construction and Automorphisms of Buildings

Mark A. Ronan

Mathematics Department, University College London, Gower Street, London, WC1E 6BT, England
e-mail: ronan@uic.edu

Introduction

This paper discusses the construction of buildings of all types: spherical, affine, hyperbolic, etc., and their automorphism groups. In some cases, including all spherical buildings and most buildings for Kac-Moody groups, the local structure determines the global structure. When this happens local automorphisms extend to global automorphisms, providing large and interesting automorphism groups, and there is a direct construction, using blueprints. In some other cases there are buildings having rather trivial automorphism groups, in which case a construction tends to allow great flexibility. To obtain interesting groups, one needs to build in additional data zboth types.

The structure of the paper is as follows. Section 1 gives a brief introduction to buildings and apartments, and section 2 goes on to spherical buildings, which is where the subject started. I explain the connection between spherical buildings and algebraic groups, and give the classification of spherical buildings. In section 3 we deal with affine buildings, which arise from groups over a field endowed with a valuation into the integers. Section 4 then explains some constructions of buildings, particularly using blueprints. A blueprint is a set of rank 2 data, usually obtained from root groups, and these in turn lead naturally to Kac-Moody groups and twin buildings, which is the topic of section 5. Twin buildings in which all rank 2 resides are spherical admit a local-to-global theorem, similar to that of spherical buildings, and this leads to a classification. Then in section 6 we discuss twin trees, and in section 7 some non-2-spherical twin buildings, involving twin trees as residues. Finally in section 8, I give a few concluding remarks and mention an important open question.

This paper originates from my lecture inaugurating the excellent conference on geometric group theory at the Indian Institute of Technology in Assam during December 2002, and I would like to thank Meenaxi Bhattacharjee for her kind invitation and her wonderful organization of the conference.

6.1. Buildings of type *W*

Throughout this paper W is a *Coxeter group*, that is to say a group generated by involutions $s_1, \ldots, s_n$, along with defining relations $(s_i s_j)^{m_{ij}} = 1$. One allows the possibility that m_{ij} is infinity. The *rank* of W is n. For any subset J of $\{1, \ldots, n\}$ the group W_J generated by the $\{s_j\}_{j \in J}$ (along with the relations for pairs of generators) embeds as a subgroup of W, so in particular W is non-trivial (see [T3], cf. [[R2];(2.1)]). If each $s_j (j \in J)$ commutes with each $s_k (k \in K)$ then the direct product $W_J \times W_K$ embeds as a subgroup of W. A Coxeter group having no direct factors W_J is called *irreducible*.

The spherical case

When W is finite the involutions s_i can be treated as orthogonal reflections in $\mathbb{R}^n$, each fixing the origin. These reflections, and their conjugates in W, create a triangulation of the $(n-1)$-sphere in which each $(n-1)$-dimensional simplex is a fundamental domain for the action of W. These fundamental domains are called *chambers*, and the triangulated sphere is the *Coxeter complex* for W. A finite Coxeter group W is called *spherical* for obvious reasons.

For example, in the rank 2 case W is a dihedral group of order $2m$, and the Coxeter complex is a $2m$-gon (thought of as the barycentric subdivision of a regular m-gon). In the rank 3 case W is either a direct product of a cyclic and a dihedral group, or it is irreducible. In the irreducible case there are two possibilities, A_3 and $B_3 = C_3$. In A_3 each chamber has dihedral angles $60°, 60°, 90°$; and in $B_3 : 45°, 60°, 90°$.

The affine case

When the involutions s_i, and their conjugates in W, can be represented as reflections in Euclidean space, not all fixing the same point, then W is called *Euclidean* or *affine*. The reflections provide a tiling of Euclidean space, called the *Coxeter complex* for W. Each maximal dimensional tile is a fundamental domain for the action of W on the Euclidean space, and is called a *chamber*.

For example in the rank 2 case, W is the infinite dihedral group, which acts on the Euclidean line tiled by segments of unit length. In the rank 3 case there are three possibilities, known as $\tilde{A}_2$, $\tilde{B}_2$ and $\tilde{G}_2$. The chambers are Euclidean triangles whose angles are: $60°, 60°, 60°$ for $\tilde{A}_2$; $45°, 45°, 90°$ for $\tilde{B}_2$; and $30°, 60°, 90°$ for $\tilde{G}_2$.

Each irreducible affine Coxeter complex is a tiling of Euclidean space by simplexes, so its rank is one greater than the dimension of the Euclidean space. In the reducible case the tiling is not by simplexes. For example, the smallest reducible affine Coxeter group has rank 4 and is a direct product of two infinite dihedral groups. Its Coxeter complex is the rectangular lattice in the plane, each rectangle being a chamber.

The hyperbolic case

When the Euclidean space of the last paragraph is replaced by hyperbolic space, then W is called *hyperbolic*. Again the fundamental domains are called *chambers*, and the reflections create a tiling of hyperbolic space, called the *Coxeter complex* for W.

There are a great many hyperbolic Coxeter groups. For example any triangle having dihedral angles that divide $180°$ and add up to less than $180°$ can be used to tile the hyperbolic plane, and there is a rank 3 Coxeter group in each case. A quite different example is obtained by tiling the hyperbolic plane with regular hyperbolic n-gons ($n \geq 5$) having dihedral angles of 90°. The chambers are n-gons, and the Coxeter group has rank n; it is generated by involutions s_i and relations $(s_i s_{i+1})^2 = 1$ as i ranges over the integers modulo n. This Coxeter complex, whose chambers are right-angled n-gons, will appear again in section 7.

Buildings and apartments

Given a Coxeter group W, a *building* of type W is a union of Coxeter complexes of that type, called the *apartments* of the building. Any two chambers lie in a common apartment. I shall not give a formal definition, but the reader is referred to my book [R2]. A building is called *spherical, affine*, etc. if its apartments are.

The codimension 1 faces of a chamber are called *panels*, and the number of panels on a chamber is called the *rank* of the building; it is the same as the rank of the Coxeter group. When the chambers are simplexes, the rank is one greater than the dimension. This is the case for all spherical and all irreducible affine buildings. Buildings of rank zero are vacuous, those of rank 1 are discrete sets of points, and those of rank 2 are bipartite graphs. A spherical building of rank 2 is called a *generalized polygon*, and a non-spherical building of rank 2 is a tree.

6.2. Spherical buildings and algebraic groups

The theory of buildings was created by Jacques Tits from his study of semi-simple Lie groups and classical projective geometry. In the early 1950's he started to examine the geometry of the exceptional simple Lie groups (those of types E_6, E_7, E_8, F_4 and G_2) from a combinatorial point of view, hoping to construct such groups over arbitrary fields. The group E_6 had been dealt with by Dickson [D] using a cubic form in 27 dimensions, but Tits wanted to deal with all the exceptional groups in a uniform manner. Then in 1955, Claude Chevalley published his famous paper [Ch] on "Chevalley groups" – analogues of complex semisimple Lie groups over arbitrary fields. Soon afterwards, variations on Chevalley's methods were published by Chevalley himself, by Steinberg [St], Tits and others. These "twisted" Chevalley groups arise over fields that are not algebraically closed, and are now subsumed under the banner of algebraic groups, where they appear as groups of rational points, as explained below.

Algebraic groups

An algebraic group is a matrix group over an algebraically closed field, and its elements are defined as the solutions to a set of polynomial equations. We are concerned here with the semisimple ones, and the group $SL_n(\mathbf{C})$ is a useful example; its elements are given by a single equation in the n^2 matrix entries, namely determinant $= 1$.

Descending from an algebraically closed field to a subfield, one uses the term *rational points* of an algebraic group to mean points having entries in the subfield. The group $SL_n(\mathbb{R})$ is one obvious way of descending from $SL_n(\mathbf{C})$, but there are others. Consider $SL_2(\mathbf{C})$ as an example. It is the set of matrices $\left(\begin{smallmatrix} a & b \\ c & d \end{smallmatrix}\right)$ satisfying the equation $ad - bc = 1$. This equation uses the four coordinates $\left(\begin{smallmatrix} 1 & 0 \\ 0 & 0 \end{smallmatrix}\right)$, $\left(\begin{smallmatrix} 0 & 1 \\ 0 & 0 \end{smallmatrix}\right)$, $\left(\begin{smallmatrix} 0 & 0 \\ 1 & 0 \end{smallmatrix}\right)$, $\left(\begin{smallmatrix} 0 & 0 \\ 0 & 1 \end{smallmatrix}\right)$, and restricting a, b, c, d to the real numbers gives $SL_2(\mathbb{R})$. Now take a different set of coordinates: $\left(\begin{smallmatrix} 1 & 0 \\ 0 & 1 \end{smallmatrix}\right)$, $\left(\begin{smallmatrix} i & 0 \\ 0 & -i \end{smallmatrix}\right)$, $\left(\begin{smallmatrix} 0 & -1 \\ 1 & 0 \end{smallmatrix}\right)$, $\left(\begin{smallmatrix} 0 & i \\ i & 0 \end{smallmatrix}\right)$. Using multiples a, b, c, d of these new coordinates, each 2-by-2 matrix is expressed in the form $\left(\begin{smallmatrix} a+bi & -c+di \\ c+di & a-bi \end{smallmatrix}\right)$, and $SL_2(\mathbf{C})$ is obtained from the polynomial condition $a^2+b^2+c^2+d^2 = 1$. Restricting a, b, c, d to the real numbers gives the group $SU_2(C)$, rather than $SL_2(\mathbb{R})$. Both groups $SL_2(\mathbb{R})$ and $SU_2(\mathbf{C})$ are obtained by descent from $\mathbf{C}$ to $\mathbb{R}$, and are called *real forms* of $SL_2(\mathbf{C})$. The first is the split form and the second is the anisotropic form. Split forms exist over any field, and Chevalley's work [Ch] showed how to obtain them. The anisotropic form always exists over the real numbers and is compact as a topological group. Over some other fields, finite ones for example, there is no anisotropic form, but there may be other non-split forms, lying so to speak between the split and anisotropic forms which are extreme cases – an example appears in due course for the group $SL_6(\mathbf{C})$.

Algebraic groups and buildings

While the theory of algebraic groups was developing, Tits came up with the idea of a building. He developed axioms that could be stated geometrically, and other axioms that could be stated group-theoretically in terms of a BN-pair (see [T4]). A BN-pair involves subgroups B and N that arise naturally in a semisimple algebraic group (or more precisely its group of rational points over some field). They also arise in other contexts that we shall meet later: B stabilizes a chamber and N stabilizes an apartment in the building. The axioms for a BN-pair apply to all types of buildings. When a BN-pair arises in the usual way from a semisimple algebraic group the building is spherical. Here is the example of type A_3, arising from the group $SL_4(K)$.

Example. Take a 4-dimensional vector space V over a field or skew field K. The chambers of the building are the nested sequences $V_1 \subset V_2 \subset V_3$ of proper subspaces, where V_i has dimension i. The other simplexes of the building, the vertices and edges, are subsequences. There are three types of vertices, corresponding to subspaces of dimensions 1, 2, and 3. To obtain an apartment in the building, pick a basis for V and take all subspaces spanned by this basis. There will be four 1-spaces (think of them as the vertices of a tetrahedron), six 2-spaces (one in the middle of each edge of the tetrahedron), and four 3-spaces (one in the middle of each face of the tetrahedron). There are twenty-four nested sequences of subspaces $V_1 \subset V_2 \subset V_3$ (six for each face of the tetrahedron): these are the chambers of the apartment.

This example is easily modified by taking V to be a vector space of dimension $n+1$, and using nested sequences of subspaces $V_1 \subset V_2 \subset \ldots \subset V_n$ as chambers. The building in this case has rank n (dimension $n-1$) and belongs to the group $SL_{n+1}(K)$. As to the BN-pair: a chamber stabilizer B is the subgroup of upper triangular matrices; and if N is taken as the full stabilizer of an apartment it comprises the subgroup of matrices having a single non-zero entry in each row and column. N contains the diagonal subgroup T (T for torus), and N/T is the Weyl group (Coxeter group) W, isomorphic to the symmetric group S_{n+1}.

Spherical buildings and rational points of algebraic groups

The connection between spherical buildings and algebraic groups is as follows. Take a semisimple algebraic group defined over a field k. As an algebraic group over the closure $\bar{k}$ of k it belongs to one of the families A_n, B_n, C_n, D_n, E_6, E_7, E_8, F_4 or G_2 (note that for buildings there is no distinction between types B_n and C_n). The rational points over k form a subgroup $G(k)$ having a BN-pair, and this gives rise to a spherical building on which $G(k)$ acts as a group of automorphisms. If the group $G(k)$ is split, in the technical sense used above, then its building has the same type as that of the algebraic group $G(\bar{k})$ over the closure of k. If $G(k)$ is not split then its building has a different type, and lower rank, from that of $G(\bar{k})$.

As an example, the algebraic group $SL_6(\mathbf{C})$ has a building of type A_5, and there are three real forms: one is the split form $SL_6(\mathbb{R})$ whose building also has type A_5; another is the compact form $SU_6(\mathbf{C})$ whose building is vacuous; and the third type is SL_3 over the quaternions whose building has type A_2 and is the flag-graph of the quaternionic projective plane. A classification of all possible rational forms of semi-simple groups is given by Tits in [T2].

Automorphisms and classification of spherical buildings

In his book [T4], Tits develops the fundamental ideas of building theory and at the same time classifies spherical buildings. Before discussing the classification one needs the elementary fact (e.g. [R2;3.10]) that a building can be split into a direct product of irreducible components, one for each direct factor of the Coxeter group (these in turn are in one-to-one correspondence with the connected components of the diagram). The classification problem therefore reduces to classifying irreducible buildings, those having a connected diagram. In the spherical case of rank at least 3 this means one of the types $A_n(n \geq 3)$, $B_n = C_n(n \geq 3)$, $D_n(n \geq 4)$, E_6, E_7, E_8, F_4. The other types of finite Coxeter groups, namely H_3 and H_4, were eliminated by Tits [T5] using the non-existence of Moufang generalized 5-gons (see (2.2) and (2.4) below).

Tits' approach to classification was to show that an irreducible spherical building of rank ≥ 3 must be determined by its local data, and must admit automorphisms taking one local datum to another. The chambers on a codimension 2 face of a spherical building Δ form a rank 2 building, referred to as a rank 2 *residue* of Δ, and the term local means the union of the rank 2 residues on a given chamber. To state the relevant "local – to – global" theorem, let c be a chamber and let $E_2(c)$ denote the subcomplex formed by chambers having a face of codimension 2 in common with c (or in other words the union of the rank 2 residues containing c). A building is *thick* if each panel lies on at least three chambers.

Theorem 6.2.1 (Tits [T4;4.1.2]). *Let c, c' be chambers in spherical buildings Δ and Δ' that are irreducible and of rank at least 3. Then any isomorphism from $E_2(c)$ to $E_2(c')$ extends to an isomorphism from Δ to Δ'.*

If Δ and Δ' are the same building then this theorem shows that a local isomorphism extends to a global automorphism. If in addition $c = c'$, then the existence of certain local automorphisms can be deduced from a slightly stronger version of (6.2.1), showing the buildings to be Moufang, a term I shall now explain.

The Moufang Condition. If α is a root (half-apartment) in a spherical building, let U_α denote the subgroup fixing α and fixing all chambers having a panel in the interior of α. It is straightforward to show, using [T4;4.1.1] that U_α must act freely on the on the set of apartments containing α. When this action is transitive the building is called *Moufang*, and U_α is called a *root group*. Using a slightly stronger version of the theorem above, namely [T4;4.16], one shows that all irreducible, spherical buildings of rank at least 3 are Moufang. A more general definition of Moufang for non-spherical buildings is given in [R2]. An important feature of the Moufang property is that it applies to the irreducible rank 2 residues.

Theorem 6.2.2. *Let Δ be an irreducible, spherical building of rank at least 3. Then every irreducible rank 2 residue is Moufang.*

A rank 2 spherical building is also called a generalized m-gon ($m \geq 3$). A root α in a generalized m-gon is a path $(x_0, \ldots, x_m)$, and U_α fixes those vertices adjacent to any of $x_1, \ldots, x_{m-1}$. For example when $m = 3$ (the case of projective planes) U_α fixes all vertices adjacent to x_1 or x_2. In terms of points and lines of the plane, one can take x_1 as a point p and x_2 as a line L on p, so a root group fixes all lines on p and all points on L. When the root group is transitive on the points of any line through p, except the line L, the plane is called Moufang, and this is the origin of the term. According to (6.2.2), in an irreducible, spherical building of rank at least 3, all residues of type A_2 are Moufang planes.

The classification of spherical buildings

The types of residues in a building are given by sub-diagrams of the diagram for the Coxeter group W, and in all irreducible, spherical cases except B_3 and F_4 there are sub-diagrams of type A_3. This implies (by a classical theorem of projective geometry – see [T4;6.7]) that the planes must be Desarguesian, rather than just Moufang. Desarguesian means they are coordinatized by field or skew-field. The classification of irreducible, spherical buildings of rank ≥ 3 is as follows.

Type $A_n (n \geq 3)$. These all arise from projective space of dimension n over a field (possibly non-commutative), and the result is a classical theorem of projective geometry.

Types $D_n (n \geq 4)$, E_6, E_7, E_8. All these diagrams have single bonds. The existence of A_3 subdiagrams implies that the planes are coordinatized by a field or skew-field, and the D_4 subdiagram implies it cannot be a skew-field (see [T4;6.12]). Thus there is one building of each of these types for each commutative field.

Type $B_n = C_n (n \geq 3)$. These are called *polar spaces* by Tits. When $n \geq 4$, there is an A_3 sub-diagram, so all the projective planes are coordinatized by a field or skew-field K. Tits then showed that the buildings are embeddable in a vector space over K, possibly infinite dimensional, and arise from algebraic forms (quadratic forms, sesquilinear forms, pseudo-quadratic forms).

When $n = 3$ there is no A_3 subdiagram and the projective planes could be non-Desarguesian, though they must be Moufang. A Moufang non-Desarguesian plane is coordinatized by an alternative division algebra, and its automorphism group arises from a rational form of E_6 (see [T1]). A B_3 building having such planes arises from a rational form of E_7. These, along with the embeddable polar spaces give a complete classification of all buildings of type $B_n (n \geq 3)$. The classification of B_n buildings, both the embeddable and non-embeddable ones is a long and technical process in [T4]. An alternative method is now available, due to the classification of Moufang polygons (rank 2 buildings) by Tits and Weiss [TW]. Using their classification of B_2 buildings (which is itself a complicated process) they can swiftly move up a rank to the B_3 case – see Chapter 40 in [*loc.cit.*].

Type F_4. There are five types. One arises from a split group of type F_4, and exists over any field. Another is an extension of the B_3 building with non-Desarguesian planes, which arises from a rational form of E_7 (see the paragraph above); its F_4 extension arises from a rational form of E_8. For the other three, see [T4;Ch.10] and cf. [R2;Ch.8].

Theorem 6.2.3. *The above list gives a classification of all irreducible spherical buildings of rank at least 3.*

For buildings of rank 2, known as generalized polygons, there is no classification unless one assumes the existence of root group automorphisms, in which case they are called Moufang (see above). Proofs published in 1979 by Tits [T6] and Weiss [W] showed that for thick Moufang m-gons, m must be 3, 4, 6 or 8. More than twenty years later Tits and Weiss [TW] completed a classification of Moufang buildings of rank 2.

Theorem 6.2.4 (see [TW; Ch.17]). *A Moufang m-gon must have m equal to 3, 4, 6 or 8 and there is a complete list in each case.*

6.3. Affine buildings and discrete valuations

The first affine buildings emerged from work of Iwahori and Matsumoto [IM] for Chevalley groups over p-adic fields. The general theory was then developed by Bruhat and Tits [BT], and a classification was finally given by Tits [T8].

Affine buildings arise from semisimple groups over fields endowed with integer valuations. Here are some examples of such fields. The rational numbers $\mathbb{Q}$ has an integer valuation $v_p(p^n \frac{a}{b}) = n$ where p is a prime not dividing a or b. This is the p-adic valuation, and the completion of $\mathbb{Q}$ with respect to this valuation is the field of p-adic numbers $\mathbb{Q}_p$. A quite different example is obtained from a field $k(t)$ of rational functions over some coefficient field k. There are many different integer valuations of

$k(t)$, one of which is given by $v\left(\frac{f}{g}\right) = deg\ g - deg\ f$, where f and g are polynomials. The completion of $k(t)$ with respect to this valuation is the field $k((t))$ of Laurent series.

Associated to any field having a discrete valuation is its *residue field*. This is obtained from the ring comprising those elements having valuation ≥ 0, and factoring out by the maximal ideal of elements having valuation ≥ 1. In the examples above, the residue field for $\mathbb{Q}$ with valuation v_p is the finite field $\mathbb{F}_p$ of p elements. Completion to $\mathbb{Q}_p$ has no effect on the residue field. For the field $k(t)$ with the valuation given above, the residue field is k.

The relationship of integer valuations to affine buildings is as follows. Let K be a field having an integer valuation with residue field k. A semisimple group over K acts on an affine building whose local structure is over the field k. For example if the group is $SL_n(K)$ and if x is a vertex of the affine building, then $\mathrm{St}(x)$ (the residue of x formed from the chambers having x as a vertex) is the spherical building for $SL_n(k)$. Thus the affine buildings for $SL_n(\mathbb{Q}_p)$ and $SL_n(\mathbb{F}_p((t)))$ are locally isomorphic, but for $n \geq 3$ they are not globally isomorphic. The fields $\mathbb{Q}_p$ and $\mathbb{F}_p((t))$ have the same residue field, and they are not the only ones; any totally ramified extension of $\mathbb{Q}_p$ has the same residue field, so there are infinitely many isomorphism classes of affine buildings of rank ≥ 3 having the same local structure. In particular Theorem 6.2.1 on spherical buildings fails for affine buildings.

The simplest example of an affine building is the tree for $SL_2(K)$ where K has residue field k. The star of each vertex in this tree is a set of edges in bijective correspondence with the rational points of the projective line over k. For example if k is the finite field $\mathbb{F}_p$ then there are $p+1$ edges per vertex. Notice that the trees for $SL_2(\mathbb{Q}_p)$ and $SL_2(\mathbb{F}_p((t)))$ are isomorphic (if all vertices of a tree have the same valency, as they do in this case, then this valency determines the tree up to isomorphism). However as soon as we move to SL_3 the affine buildings for $\mathbb{Q}_p$ and $\mathbb{F}_p((t))$ fail to be isomorphic.

The building at infinity

Since the local-to-global theorem for spherical buildings (Theorem 6.2.1) fails for affine buildings, one might expect them to be difficult to classify, but one important fact helps immensely. An affine building of dimension n has a spherical building of dimension $n-1$ at infinity. Each affine apartment has a spherical apartment at infinity, and all apartments of the spherical building at infinity arise from affine apartments in this way. The simplest example of an affine building is a tree; its building at infinity is its set of ends.

The building at infinity arose in the work of Bruhat and Tits [BT], and was then used by Tits [T8] to effect a classification. The idea is this. If an irreducible affine building has dimension at least three, then its spherical building at infinity is irreducible and has dimension at least 2, so we can apply the classification of Theorem 6.2.3. The problem, by no means an easy one, is to reconstruct the valuation of the field using the affine structure. This is done by Tits [T8] who makes particular use of trees (attached to panels of the building at infinity). He proves:

Theorem 6.3.1 (T8; sect.14). *Every irreducible affine building of dimension at least 3 has at infinity a spherical building of dimension at least 2 whose base field is complete with respect to an integer valuation. This valuation uniquely determines the affine building.*

Corollary 6.3.2 (T8; sect.15). *The locally finite affine buildings that are irreducible and of dimension $d \geq 3$ are the affine buildings of simple algebraic groups of rank d over locally compact fields that are non-discrete and totally discontinuous.*

See Tits [*loc.cit.*] for further details about the valuations and their relationship to the building structure.

In closing this section, I should like to point out what happens when the field is not complete with respect to the valuation. Let K be such a field and $\hat{K}$ its completion; for example K could be $k(t)$ in which case $\hat{K}$ would be $k((t))$. The affine buildings, for instance those for $SL_n(K)$ and $SL_n(\hat{K})$, are isomorphic as simplicial complexes. The difference occurs with the system of apartments. In the complete case all Coxeter subcomplexes of type W are apartments, but in the non-complete case a certain subset of apartments, called a *system of apartments* is used. Going to infinity only in these apartments gives the spherical building for the group over K rather than the complete spherical building for the group over $\hat{K}$.

For example for SL_2 the affine building is a tree. Any two ends of this tree span an apartment, and the complete building at infinity is the set of all ends; it is in bijective correspondence with the projective line over the complete field $\hat{K}$. For the lesser field K one has the same tree but a lesser set of ends, in bijective correspondence with the points of the projective line over K. Any two of these ends span an apartment in the system of apartments for $SL_2(K)$.

Tits' work [T8] deals extensively with systems of apartments, and the appropriate definitions can be found in section 1 of that paper.

6.4. Constructions

Buildings that do not arise from groups with a BN-pair have been constructed, sometimes with a great deal of freedom. Many of those that do arise from groups have also been constructed independently of the groups, and this section will deal with both types.

Free constructions

The first "free" construction of buildings was given by Tits for generalized m-gons (rank 2 spherical buildings) – see [T7], and [R2;Ch.3,Ex.21]. Tits showed that generalized m-gons exist for any $m > 1$. Moreover the freedom of the construction showed that they were not susceptible of classification without additional assumptions.

One additional assumption is the Moufang condition on the existence of root group automorphisms (see Sect. 2). Another quite different assumption is finiteness, in which case W. Feit and G. Higman [FH] proved that generalized m-gons with at least three edges per vertex could exist only for $m = 2, 3, 4, 6$ or 8. In the cases $m = 6$ or 8 the only known finite examples are those arising from algebraic groups, so there may well be a classification theorem in these cases – see [VM] for more details on finite generalized m-gons.

Let us now turn to affine buildings of rank 3 (dimension 2). Their rank is insufficient for the classification theorem (6.3.1), and in [R1] I showed that a free construction exists, even in the locally finite case. The method applies to any building whose rank 3 residues are non-spherical. It works by starting at one chamber and working outwards. The rank 2 residues can be chosen independently, the only restriction being the obvious one that if two rank 2 residues have a rank 1 residue in common then they confer the same valency on this rank 1 residue. Assuming valency at least 3 this implies that two rank 1 residues of the same type have the same valency.

In order to state a theorem, let us say that a building has parameters $(q_i), i \in I$, if every rank 1 residue of type i has $q_i + 1$ chambers; we allow the possibility that q_i may be infinite. This means that every rank 2 residue of type $\{i, j\}$ is a generalized m_{ij}-gon with parameters (q_i, q_j); if m_{ij} is infinity this means it is a tree. The fact that every rank 3 residue is non-spherical means that $\frac{1}{m_{ij}} + \frac{1}{m_{jk}} + \frac{1}{m_{ki}} \leq 1$ for every set of three types $i, j, k \in I$. Define (q_i) to be an *admissible* set of parameters if for each i, j there exists a generalized m_{ij}-gon having parameters (q_i, q_j). For example if $m_{ij} = 3$ then $q_i = q_j$ and in all known finite examples q_i is a prime power.

Theorem 6.4.1 (R1). *Let (q_i) be an admissible set of parameters, and W a Coxeter group having no rank 3 spherical subgroups. Then there exists a building of type W with these parameters, whose rank 2 residues of type $\{i, j\}$ range through any desired set of generalized mij-gons having parameters (q_i, q_j).*

This theorem, of course does not apply to irreducible, spherical buildings of rank at least 3, and for that we need a more sophisticated approach. This approach, developed by Tits and the author uses the concept of a “blueprint”.

Blueprint constructions

If a building contains irreducible spherical rank 3 residues (namely of type A_3 or B_3) then its local structure is more restricted than in the construction above. For example in an A_3 or B_3 residue two rank 2 sub-residues intersecting in a rank 1 residue must give compatible structures on that intersection. This compatibility problem led the author and J.Tits [RT1] to a new method of construction that, although not as free as the earlier one in [R1], applies to buildings of all types. Here is how it works.

Let I be the indexing set for the fundamental roots, or equivalently for the different types of panels in the building. Attach to each i in I a set U_i. The idea is to use the sets U_i, along with the Coxeter group, to parametrize the building, starting with one chamber b and working outwards. A *gallery* from b to another chamber x is a sequence of chambers $b = x_0, x_1, \ldots, x_d = x$, each one adjacent to the next; we say it has *type* $(i_1, \ldots, i_d)$ if x_{j-1} is i_j-adjacent to x_j. Using the sets U_i we then specify x by a sequence $(u_{i_1}, \ldots, u_{i_d})$, where $u_j \in U_j$, and $(i_1, \ldots, i_d)$ is the type of a minimal gallery from b to x.

If b is joined to x by minimal galleries of different types then it is necessary to know when sequences of two different types yield the same chamber. Fortunately, if we fix the end chambers b and x, one minimal gallery can be transformed into another by a sequence of alterations (called *elementary homotopies*) in rank 2 residues (see [T7;(3.4)]). The datum consisting of the sets U_i along with these rank 2 homotopies, for each type of rank 2 residue, is called a *blueprint*. A building that can be parametrized in accordance with a blueprint is said to *conform* to that blueprint, and a blueprint is called *realisable* if there exists a building conforming to it. The secret in producing realisable blueprints is to let the U_i be root groups for the fundamental roots – see [RT1] for details. Clearly any realisable blueprint must be realisable in all rank 3 spherical residues, and these are the residues that formed an obstruction to the previous construction. What is surprising is that the irreducible rank 3 spherical residues are the only obstruction.

Theorem 6.4.2 (RT1; Theorem 1). *A blueprint is realisable if its restriction to each rank 3 spherical subdiagram is realisable. In this case there is a unique building which conforms to it.*

[Note: the statement in [RT1] inadvertently omits the word spherical].

The reason the rank 3 spherical restriction is sufficient can be explained in terms of galleries as follows. A sequence of rank 2 homotopies may yield a gallery having the same type as the initial gallery. But the initial sequence $(u_{i_1}, \ldots, u_{i_d})$ and the final sequence $(v_{i_1}, \ldots, v_{i_d})$ must yield the same end chamber, so $u_{i_j} = v_{i_j}$. To ensure this happens, it suffices to check such self-homotopies in rank 3 spherical residues because of a theorem for Coxeter complexes given by Tits in [T7; Prop. 4].

Foundations

In order to avoid explicit discussion of blueprints, the authors of [RT1] defined a *foundation* to be an amalgam of rank 2 residues containing a given chamber. It is said to *support* a building if there exists a building and a chamber c in that building such that $E_2(c)$ – the union of the rank 2 residues containing c – is isomorphic to the foundation concerned. Rank 3 foundations that are reducible always support a building that is a direct product of a rank 2 residue and a rank 1 residue, so the only issue in the rank 3 spherical case is the residues of types A_3 and $B_3 = C_3$. The point of using foundations is the following powerful existence theorem.

Theorem 6.4.3 (RT1; Theorem 2). *Let E be a foundation. If each rank 3 residue of type A_3 or B_3 supports a building, then E supports a building.*

This theorem is proved by constructing a realisable blueprint, and hence a building conforming to it. The question then is which buildings are obtained in this way, conforming to a blueprint.

In the spherical case *all* buildings conform to a blueprint. In particular the theorems above prove the existence of spherical buildings, such as those of type E_8, that were previously only known from the existence of the Chevalley groups. There is one building of type E_8 for each commutative field. The commutativity of the field stems from the existence of a D_4 subdiagram – see [RT1; Sect. 5] for an explanation of this in terms of blueprints.

In the affine case the buildings conforming to a blueprint are those for groups over a power series field, rather than a p-adic field. The p-adic affine buildings do not conform to blueprints, apart from an exceptional case where the residue field is the field of two elements. In other cases of buildings, such as hyperbolic ones, those conforming to a blueprint include all buildings attached to Kac-Moody groups. This brings us to the next section.

6.5. Twin buildings and Kac-Moody groups

Kac-Moody Lie algebras generalize semisimple Lie algebras, allowing infinite Weyl groups (Coxeter groups) W, and they lead to Kac-Moody groups in the same way that semi-simple Lie algebras lead to Chevalley groups. When W is a finite group (the spherical case) the algebra is finite dimensional (and semisimple); when W is infinite the algebra is infinite dimensional. The buildings have the same rank as W and are finite dimensional.

The first method of attaching groups to these algebras was given by Moody and Teo [MT]. Later on, Tits wrote a detailed study of Kac-Moody groups [T9] generating the groups using root groups U_α where α ranges over the set of roots of the Coxeter complex for W (recall that a root is half of a Coxeter complex, on either side of a reflecting hyperplane). These root groups, each isomorphic to the additive group of k, generate a group G. If N denotes the subgroup of G normalizing the set of root groups, and T denotes the subgroup normalizing each individual root group, then as in the case of Chevalley groups, N/T is isomorphic to the Coxeter group W. The subgroup T, called a torus, is isomorphic to a direct product of copies of the multiplicative group of k.

If c denotes a chamber of the Coxeter complex, the positive roots are those containing c, and the negative roots are those not containing c. The positive and negative root groups generate groups U_+ and U_-, which along with the torus generate subgroups B_+ and B_- respectively. The pairs (B_+, N) and (B_-, N) are BN-pairs for G.

A very important difference between the spherical case and the non-spherical case is that B_+ and B_- are conjugate in the spherical case, but not otherwise. The reason is simple. The element conjugating B_+ to B_- in a Chevalley group is the longest word in W, but when W is infinite there is no longest word.

The BN-pairs (B_+, N) and (B_-, N) generate two buildings Δ_+ and Δ_- whose rank is the rank of W. There is a W-valued "codistance" between chambers of Δ_+ and Δ_-, providing (Δ_+, Δ_-) with the structure of a twin building – see [T10] for the twin building axioms. When W is finite (the spherical case) then $\Delta_+ = \Delta_-$, and the codistance is the usual W-valued distance pre-multiplied by the longest word in W. A spherical building is twinned with itself.

Two chambers, one in one building and one in the other, are said to be opposite one another when the codistance between them is the identity element of W (and two vertices are *opposite* if they belong to *opposite* chambers and have the same type).

Each twin building (Δ_+, Δ_-) contains *twin apartments*, which are pairs (A_+, A_-) of apartments one in each building, such that each chamber of A_+ is opposite a unique chamber of A_-. In the non-spherical case a twin apartment can be thought of as having its two components on opposite sides of a sphere, so that a chamber in one component of a twin apartment is opposite its antipode in the other component.

Example. The group SL_n over a ring of Laurent polynomials $k[t, t^{-1}]$ is a Kac-Moody group of affine type and acts naturally on a twin building (Δ_+, Δ_-). Both Δ_+ and Δ_+ are buildings for $SL_n(k(t))$ but using different valuations v_+ and v_-, where $v_+(t) = 1$ and $v_-(t^{-1}) = 1$. The subgroup $SL_n(k[t])$ fixes a vertex x_+ of Δ_+, and $SL_n(k[t^{-1}])$ fixes an opposite vertex x_- of Δ_-. The stabilizer of the pair (x_+, x_-) is $SL_n(k)$.

When k is a finite field, $SL_n(k[t])$ acts discretely on Δ_-, and similarly with $SL_n(k[t^{-1}])$ acting on Δ_+. This makes affine twin buildings useful in the study of arithmetic groups over a ring of Laurent polynomials – see [A] for more details.

2-spherical twin buildings

Twin buildings can be viewed as generalizations of spherical buildings, and when they are *2-spherical* (meaning that all rank 2 residues are spherical) Tits' methods for spherical buildings can be used to prove that the rank 2 residues must be Moufang A complete proof is given in [[R3]; Theorem 3].

Theorem 6.5.1 (R3). *Let Δ be a 2-spherical twin building with no direct factors of rank 2. Then every residue of type $\{i, j\}$ for which $m_{ij} \geq 3$ is Moufang. In particular m_{ij} is 2,3,4,6 or 8 for all $\{i, j\}$.*

Moreover one can show that the local structure usually determines the global structure, as in theorem 6.2.1. Recall that $E_2(c)$ means the union of the rank 2 residues containing the chamber c, referred to as the *local structure*. In the following theorem an *isometry* is a map preserving distances and codistances.

Theorem 6.5.2 (MR). *Let c and c' be pairs of opposite chambers in twin buildings Δ and Δ' respectively, and assume Δ has no residues of type $Sp_4(2)$, $G_2(2)$, $G_2(3)$ or ${}^2F_4(2)$. Then any isometry from $E_2(c)$ to $E_2(c')$ extends to an isometry from Δ to Δ'. In particular the local structure of Δ determines its global structure.*

The reason for excluding certain residues over the fields of 2 or 3 elements is that in these cases the set of chambers opposite a given chamber is not connected. In all other cases the set of chambers opposite a given chamber is connected, and the building is said to satisfy condition (co).

Using this Local-to-Global theorem the classification of 2-spherical twin buildings has been achieved in the following way (see [M1] and [M2] for further details).

Theorem 6.5.3 (M2). *Let Δ be an irreducible, 2-spherical twin building satisfying (co). Then one of the following holds:*

1. *Δ is associated with a classical group over a division ring that is infinite dimensional over its centre. In this case Δ is of spherical or of affine type.*
2. *Δ can be obtained from the class of twin buildings associated with Kac-Moody groups over algebraically closed fields by a sequence of constructions (CS), (FP), (L), (FP), (FP).*

(CS) refers to Convex Sub-buildings of the same type, obtained by descending to a suitable subfield.
(L) refers to Limits, such as moving from finite dimensional to infinite dimensional forms of finite Witt index.
(FP) refers to Fixed Points.

All three procedures occur in the spherical case when descending from a building over an algebraically closed field, but they are only needed in the sequence (CS), (FP), (L). In the non-spherical case there are added complications because the diagram can contain circuits, and the fixed point procedure can involve descent from a building of infinite rank to one of finite rank (see [M2] for more details).

6.6. Twin trees

Rank 2, non-spherical twin buildings are twin trees. Their definition can be given very simply in terms of an integer codistance between vertices, rather than using the more general definition of a twin building in terms of a W-valued codistance between chambers. Before stating this definition, note that a single tree can be defined in terms of a set T of vertices along with a non-negative integer $\text{dist}(x, y) = \text{dist}(y, x)$, called the *distance*, between any two of them. One has $\text{dist}(x, y) = 0$ if and only if $x = y$, and when $\text{dist}(x, y) = 1$ one says that x and y are *neighbours*. With this distance function, T is a tree precisely when the following condition is satisfied:

if $\text{dist}(x, y) = n$ *and if* y' *is a neighbour of* y*, then* $\text{dist}(x, y') = n \pm 1$*; furthermore if* $n > 0$*, then* -1 *occurs for a unique such neighbour* y'.

A twin tree is a pair of trees, with each vertex having at least two neighbours, and with a non-negative integer $\text{codist}(x, y) = \text{codist}(y, x)$, called the *codistance*, between vertices of one and vertices of the other. This codistance satisfies the following condition:

if $\text{codist}(x, y) = n$ *and if* y' *is adjacent to* y*, then* $\text{codist}(x, y') = n \pm 1$*; furthermore if* $n > 0$*, then* $+1$ *occurs for a unique such neighbour* y'.

Vertices x and y are opposite when $\text{codist}(x, y) = 0$. Opposite vertices necessarily have the same valency, and when the trees are thick (meaning that every vertex has valency at least three), then vertices in the same tree at even distance from one another, or vertices in opposite trees at even codistance from one another, have the same valency. In this sense a twin tree is *semi-homogenous*.

Example. In the twin tree for the group SL_2 over a ring of Laurent polynomials $\mathbb{F}_q[t, t^{-1}]$, where $\mathbb{F}_q$ is the field having q elements, each vertex has valency $q + 1$.

Ends and apartments

For a tree in which each vertex has valency at least two, an *apartment* is a path infinite in both directions. A *half-apartment* is a path infinite in one direction. Two half-apartments are said to have the same *end* if their intersection is a half-apartment. Having the same end is an equivalence relation on the set of half-apartments, and the equivalence classes are called the *ends* of the tree. Each apartment has two ends, and any two ends a and b span a unique apartment, denoted (ab).

A *twin apartment* of (T_+, T_-) is a pair of apartments A_+ in T_+ and A_- in T_- having the property that every vertex of one is opposite a unique vertex of the other. The set of ends for the twin apartments are called the ends of the twinning. This is a proper subset of the set of all ends of either tree. For example if the vertices have finite valency at least 3, then the *ends of the twinning* form only a countable subset in the uncountable set of all ends.

There is a natural bijective correspondence between the ends of the twinning for T_+ and those for T_-, so one can talk of the *ends of the twinning* without specifying to which tree they belong. For a twin apartment (A_+, A_-) the ends a and b of A_+ are the same as those of A_-; i.e. the apartment (ab) spanned by a and b is A_+ in one tree and A_- in the other. Given two ends of the twinning, one can ask whether they necessarily span a twin apartment, but the answer is no. Only some pairs have this property, and knowing those that do is equivalent to knowing the codistance function and hence knowing the twinning.

Proposition 6.6.1 (RT1;3.6). *In a twin tree having at least three edges per vertex the set of twin apartments uniquely determines the twinning.*

Moufang twin trees and Kac–Moody groups

Rank 2 Kac-Moody Lie algebras over a field k are given by a generalized Cartan matrix $\left(\begin{smallmatrix} 2 & a \\ b & 2 \end{smallmatrix}\right)$, where a and b are negative integers. When $ab = 1, 2$ or 3 the algebras are finite dimensional, giving rise to Chevalley groups of types A_2, $B_2 = C_2$, and G_2, respectively. Their buildings are generalized m-gons for $m = 3, 4$, and 6. When $ab \geq 4$ the algebra is infinite dimensional and the buildings are twin trees. In the $SL_2(k[t, t^{-1}])$ example above, $a = b = -2$.

These "Kac–Moody" twin trees are Moufang in the sense that they admit root groups (see [RT2] for the definition of root groups for twin trees), but they are by no means the only Moufang examples. In the case of valency 3 (three edges per vertex), Tits [T11] has produced uncountably many Moufang twin trees.

Constructing twin trees

In all the Kac-Moody examples the number of edges per vertex is $q + 1$ where q is a prime power or an infinite cardinal number. An obvious question is whether examples can exist for any q, and the answer is yes, as the following theorem makes clear.

Theorem 6.6.2 (RT3;8.2). *A thick semi-homogeneous tree whose set of vertices has cardinality α admits 2^α isomorphism classes of twinnings, and among these 2^α have trivial automorphism group.*

The method of proof is to examine a partial twinning between a tree and a ball of radius 2 around a given vertex. This partial twinning is called a *uniform 2-ball.* In [RT3] we show two main things: (1) that a uniform 2-ball can have a wild structure, and (2) that the uniform 2-balls at different vertices of a tree twinned with T can be independent of one another. The point is that one can amalgamate uniform 2-balls in a random way to obtain a twin tree.

6.7. Some twin buildings that are not 2-spherical

I shall now explain some recent work, by B. Rémy and the author, using Moufang twin trees to construct hyperbolic buildings that are Moufang. A Moufang building has a twin, and one can therefore fix a vertex in one half of the twin to obtain a group that acts on the other half. When the root groups are finite (so the buildings are locally finite) this group acts as a lattice on the other building.

The starting point is an apartment which is a tiling of the hyperbolic plane $\mathbb{H}^2$ by equilateral right-angled m-gons. Pick one such m-gon c and label its edges π_i in a natural cyclic order as i ranges over the integers mod m. Each π_i lies on a unique geodesic, and we let s_i denote the reflection of $\mathbb{H}^2$ across this geodesic. The reflections s_i and s_{i+1} commute, and the group W generated by the s_i, along with the relations $(s_i s_{i+1})^2 = 1$, as i ranges over the integers modulo n, is the Coxeter group for this apartment.

Each reflection (conjugate of one of the s_i) fixes a wall, which is a geodesic of $\mathbb{H}^2$, and in this wall all panels have the same type i. One therefore talks of *walls* of *type* i. A root of type i is a half-apartment bounded by a wall of type i. The roots containing c are called the *positive roots*; those not containing c are the *negative roots*. Before going further it is worth quoting an existence and uniqueness theorem for buildings having apartments of this type.

Theorem 6.7.1 ([GP] and [Bo]). *Given a set of cardinals q_i for each i in $\mathbb{Z}/m$, there exists a unique hyperbolic building whose apartments are tilings of $\mathbb{H}^2$, as above, such that the cardinality of the set of chambers on a panel of type i is $q_i + 1$.*

Let $\Delta(m, (q_i))$ denote the building of (6.7.1).

Remark 1. The rank 2 residues of $\Delta(m, (q_i))$ are of two kinds. Those of type $\{i, i+1\}$ are complete bipartite graphs, but those of type $\{i, j\}$, where $j \neq i \pm 1$, are trees. These trees are semi-homogeneous with valencies $q_i + 1$ and $q_j + 1$.

Remark 2. When the q_i are all equal to a given prime power q, the building in (6.7.1) comes from a (non-unique) Kac-Moody group over the field of order q.

Now for each i choose a field k_i, and to each root α of type i attach a group U_α isomorphic to the additive group of k_i. For each w in W, let Φ_w denote the set of roots containing c but not wc, and define U_w as the direct product of the groups U_α for α in Φ_w. Treating W as a partially ordered set, via the Bruhat ordering, U_w is a subgroup of U_v whenever $w \leq v$. We then define U_+ as the inductive limit of the U_w for w in W.

For each i let T_i denote the multiplicative group of the field k_i, and define the *torus* T to be the direct product of the T_i. The group T_i acts in a natural way on each root group of type i, and centralizes those of type $\neq i$. Thus the torus T acts on U_+, and we define the (positive) Borel subgroup B_+ as a semi-direct product of U_+ and T.

Let $\alpha = \alpha_i$ be the root containing c and having π_i on its boundary, and let $-\alpha$ be the opposite of α. The root groups U_α and $U_{-\alpha}$ can be identified with root groups generating $SL_2(k_i)$, and for each i the *Levi factor* L_i of type i is the direct product of $SL_2(k_i)$ and the groups T_j for $j \neq i$. One then generates a minimal parabolic subgroup P_i of type i using L_i and U_+. These minimal parabolics contain B_+ and can be amalgamated to form a group Λ – see [RR] for details. This group Λ acts on a twin building.

Theorem 6.7.2 (RR). *When each q_i is the cardinality of a field k_i, then the building $\Delta(m, (q_i))$ above belongs to a Moufang twinning having the group Λ, defined above, as an automorphism group.*

Remark 1. The rank 2 residues of type $\{i, j\}$, where $j \neq i \pm 1$, are trees that are semi-homogeneous with valencies $q_i + 1$ and $q_j + 1$. In this twinning two such residues that are opposite form a twin tree whose automorphism group induces $SL_2(k_i)$ on the star of vertices of one type, and $SL_2(k_j)$ on the star of vertices of the other type.

Remark 2. A semi-homogeneous tree has many different twinnings (see sect. 6), and the same is presumably true for the building $\Delta(m, (q_i))$. Some of these will be Moufang, but I expect that "most" will not.

I want to end this section with a further result from [RR] showing that the use of fields of different characteristics leads to non-linearity of the group Λ.

Theorem 6.7.3 (RR). *When the q_i are not all powers of the same prime number, then the group Λ is not linear over any field.*

6.8. Concluding remarks

Here is a quick summary of results discussed in this paper. All buildings in this section will be irreducible.

Classification theorems are available in the following cases:

(i) Spherical buildings of rank ≥ 3 – see (6.2.3).
(ii) Spherical buildings of rank 2 that are Moufang – see (6.2.4).
(iii) Affine buildings of rank ≥ 4 – see (6.3.1).
(iv) Twin buildings of rank ≥ 3 that are 2-spherical, with some very minor exceptions – see (6.5.1).

All these buildings have large automorphism groups. Buildings having a looser structure may not admit such large groups, and could be rigid. Such buildings tend to admit wild constructions, and that is certainly true in the following cases:

(i) Buildings having no rank 3 spherical residues – see (6.4.1).
(ii) Twin trees – see (6.6.2).

Presumably many non-2-spherical twin buildings could be added to this list. A construction of some Moufang, hyperbolic examples of arbitrarily high rank that do not arise from Kac-Moody groups was given in section 7.

Many classes of buildings can be constructed from a blueprint, as described in section 4. Those conforming to a blueprint include:

(i) All spherical buildings.
(ii) All affine buildings arising from groups over power series fields.
(iii) All buildings arising from Kac-Moody groups.

In all these cases the buildings admit a twinning. To what extent a building that conforms to a blueprint must be twinnable is uncertain. By (6.4.1) there are 2-spherical buildings having non-Moufang rank 2 residues and no rank 3 spherical residues. Among such buildings, some can certainly be constructed using blueprints, yet they cannot be twinnable, as this would contradict (6.5.1). I would like to end by posing the following very difficult question.

Question. If all the rank 3 residues of a building are spherical, what additional conditions will ensure that it conforms to a blueprint?

The answer has to be that *some* extra condition is required, because the p-adic affine buildings do not conform to blueprints. Is it sufficient to assume the building is non-affine? I rest my question.

References

[A] P. Abramenko, Twin Buildings and Applications to S-Arithmetic Groups, Lecture Notes in Mathematics 1641, Springer-Verlag (1996).

[Bo] M. Bourdon, Sur les immeubles fuchsiens et leur type de quasi-isometrie, *Erg. Thy. and Dyn. Sys.*, **20** (2000) 343–364.

[BT] F. Bruhat and J. Tits, Groupes réductifs sur un corps local, I. Données radicielles values, *Publ. Math. I.H.E.S.*, **41** (1972) 5–252.

[Ch] C. Chevalley, Sur certains groupes simples, *Tohoku Math. J.*, **(2)7** (1955) 14–66.

[D] L. Dickson, Linear Groups, originally published 1901, reprinted Dover, New York (1958).

[FH] W. Feit and G. Higman, The nonexistence of certain generalized polygons, *J. Algebra*, **1** (1964) 114–131.

[GP] D. Gaboriau and F. Paulin, Sur les immeubles hyperboliques, *Geo. Ded.*, **88** (2001) 153–197.

[M1] B. Mühlherr, Locally split and locally finite twin buildings of 2-spherical type, *J. reine angew. Math.*, **511** (1999) 119–143

[M2] B. Mühlherr, Twin Buildings, in Tits Buildings and the Model Theory of Groups (ed. K. Tent), *London Maths Soc. Lecture Notes*, **291** (2002) 103–117.

[MR] B. Mühlherr and M. Ronan, Local to Global Structure in Twin Buildings, *Inventiones Math.*, **122** (1995) 71–81.

[MT] R. Moody and K. Teo, Tits' systems with crystallographic Weyl groups, *J. Algebra*, **21** (1972) 178–190.

[R1] M. Ronan, A Construction of Buildings with no Rank 3 Residues of Spherical Type, in Lecture Notes in Math. 1181 (Buildings and the Geometry of Diagrams, Como 1984) Springer-Verlag (1986) 242–248.

[R2] M. Ronan, Lectures on Buildings, Academic Press 1989. (A reprint with additional material will be published by the University of Chicago Press in 2009.)

[R3] M. Ronan, Local Isometries of Twin Buildings, *Math. Zeitschrift*, **234** (2000) 435–455.

[RR] B. Rémy and M. Ronan, Topological groups of Kac-Moody type, Fuchsian twinnings and their lattices, preprint.

[RT1] M. Ronan and J. Tits, Building Buildings, *Math. Annalen*, **278** (1987) 291–306.

[RT2] M. Ronan and J. Tits, Twin Trees I, *Inventiones Math.*, **116** (1994) 463–479.

[RT3] M. Ronan and J. Tits, Twin Trees II, *Israel J. Math.*, **109** (1999) 349–377.

[S] J.-P. Serre, *Arbres, Amalgames, SL2*; Astérisque **46** (1977). English version: Trees, Springer-Verlag 1980.

[St] R. Steinberg, Variations on a theme of Chevalley, *Pacific J. Math.*, **9** (1959) 875–891.

[T1] J. Tits, Le plan projectif des octaves et les groupes de Lie exceptionelles, *Acad. Roy. Belg.*, **39** (1953) 309–329.

[T2] J. Tits, Classification of algebraic semi-simple groups, in Proc. Symp. Pure Math. Vol. 9 (Algebraic Groups and Discontinuous Subgroups, Boulder 1965), 33–62, Am. Math. Soc. (1966).

[T3] J. Tits, Le problème de mots dans les groupes de Coxeter, *Ist. Naz. Alta Mat., Symposia Math.*, **1** (1968) 175–185.

[T4] J. Tits, Buildings of Spherical Type and Finite BN-Pairs, Lecture Notes in Mathematics, **386**, Springer Verlag (1974).

[T5] J. Tits, Endliche Spiegelungsgruppen, die als Weylgruppen auftreten, *Inventiones Math.*, **45** (1977) 283–295.

[T6] J. Tits, Non-existence de certains polygones généralisés, I, II, *Inventiones Math.*, **36** (1976) 275–284; **51** (1979) 267–269.

[T7] J. Tits, A Local Approach to Buildings, In: Davis C. et al (eds.), The Geometric Vein-The Coxeter Festschrift, Springer Verlag, Berlin (1981) 519–547.

[T8] J. Tits, Immeubles de type affine, in: Buildings and the Geometry of Diagrams (Como 1984), Lecture Notes in Mathematics **1181** (159–190), Springer Verlag 1986.

[T9] J. Tits, Uniqueness and Presentation of Kac-Moody Groups over Fields, *J. Algebra* **105** (1987) 542–573.

[T10] J. Tits, Twin Buildings and Groups of Kac-Moody Type, In: Liebeck and Saxl (eds.), Groups, Combinatorics and Geometry (Durham 1990), London Math. Soc. Lecture Note Ser. **165** (249–286), Cambridge Univ. Press (1992).

[T11] J. Tits, private communication.

[TW] J. Tits and R. Weiss, Moufang Polygons, Springer Monographs in Mathematics (2001).

[VM] H. Van Maldeghem, Generalized Polygons, Monographs in Mathematics 93, Birkhäuser Verlag (1998).

[W] R. Weiss, The Nonexistence of certain Moufang Polygons, *Inventiones Math.*, **51** (1979) 261–266.

Proc. Inst. Wrk.shps. Int. Conf. – Essays in Geometric Group Theory No. 9, 2009, pp. 145–159.

Zeta Functions of Nilpotent Groups – Singular Pfaffians

Christopher Voll

School of Mathematics, University of Southampton, University Road, Southampton
SO17 1BJ, United Kingdom

Abstract. The local normal zeta functions of a finitely generated, torsion-free nilpotent group G of class 2 depend on the geometry of the Pfaffian hypersurface associated to the bilinear form induced by taking commutators in G. The smallest examples of zeta functions which are not finitely uniform arise from groups whose associated Pfaffian hypersurfaces are plane curves. In this paper we study groups whose Pfaffians define singular curves, illustrating that the local normal zeta functions may indeed invoke *all* the degeneracy loci of the Pfaffian.

7.1. Introduction and statement of results

In [8] Grunewald, Segal and Smith defined the concept of the zeta function of a group. For a finitely generated, torsion-free nilpotent group G – a $\mathfrak{T}$-group, in short – they set

$$a_n^{\triangleleft} := |\{H \triangleleft G|\ |G : H| = n\}|.$$

The group's *normal zeta function* is the formal Dirichlet series

$$\zeta_G^{\triangleleft}(s) := \sum_{n=1}^{\infty} a_n^{\triangleleft} n^{-s}.$$

Here s is a complex variable, and $\zeta_G^{\triangleleft}(s)$ defines an analytic function in some right half-plane $\{s \in \mathbb{C}|\ \mathfrak{Re}(s) > \alpha\}$ for some $\alpha \in \mathbb{Q}$ ([7]). Similarly, the *local* normal zeta function of G at the prime p counts normal subgroups of p-power index in G:

$$\zeta_{G,p}^{\triangleleft}(s) := \sum_{n=0}^{\infty} a_{p^n}^{\triangleleft} p^{-ns}.$$

As G is nilpotent, the local and global zeta functions are related by an *Euler product decomposition*

$$\zeta_G^{\triangleleft}(s) = \prod_{p \text{ prime}} \zeta_{G,p}^{\triangleleft}(s).$$

The local normal zeta functions are rational functions in p^{-s} ([8]). The global normal zeta function is called *finitely uniform* if there are finitely many rational functions $W_1(X,Y), \dots, W_n(X,Y) \in \mathbb{Q}(X,Y)$ such that for every prime p there is an r such that $\zeta_{G,p}^{\triangleleft}(s) = W_r(p, p^{-s})$, and *uniform* if it is finitely uniform with $n = 1$. (Finite) uniformity is not typical for zeta functions of groups, however. The question of the dependence of the local factors on the prime p was linked to the classical problem of counting points on varieties mod p by du Sautoy and Grunewald. In [7] they give a presentation of the local zeta functions as finite sums of products of rational functions in p and p^{-s} (which we call the *uniform components* of the presentation) and functions counting the number of $\mathbb{F}_p$-rational points of certain (boolean combinations of) algebraic varieties which are irreducible, smooth, and intersect normally. For an arbitrary collection of such varieties one would not expect these functions to be polynomials in p. The description of varieties arising in this context has been an ongoing project ever since. We refer the reader to [5], [4] and [10] for further background on zeta functions of groups.

The first example of a group whose zeta function is not finitely uniform was given by du Sautoy ([2],[3]). The factors of the normal zeta function of this class two nilpotent group depend on the number of $\mathbb{F}_p$-points of the elliptic curve

$$E = (y^2 + x^3 - Dx = 0), \quad D \in \mathbb{N}.$$

He defined a nilpotent group

$$G(E) = \langle x_1, \ldots, x_6, y_1, y_2, y_3 | \, [x_i, x_j] = M(\mathbf{y})_{ij} \rangle, \qquad (7.1.1)$$

where $M(\mathbf{y}) := \left(\begin{smallmatrix} 0 & R(\mathbf{y}) \\ -R(\mathbf{y})^t & 0 \end{smallmatrix} \right)$, and $R(\mathbf{y}) = \begin{pmatrix} Dy_3 & y_1 & y_2 \\ y_1 & y_3 & 0 \\ y_2 & 0 & y_1 \end{pmatrix}$. (Here we adopt additive notation for words in y_i.) The curve E appears as the *Pfaffian hypersurface*

$$\mathcal{P}_G := (\sqrt{\det(M(\mathbf{y}))} = 0)$$

in $\mathbb{P}^2(\mathbb{Q})$ associated to this group. In [12] we proved

Theorem 7.1.1. *For almost all primes p*

$$\zeta^{\triangleleft}_{G(E),p}(s) = W_1(p, p^{-s}) + |E(\mathbb{F}_p)| W_2(p, p^{-s}), \qquad (7.1.2)$$

where $|E(\mathbb{F}_p)|$ is the number of $\mathbb{F}_p$-rational points of E and

$$W_1(X, Y) = \frac{(1 + X^6Y^7 + X^7Y^7 + X^{12}Y^8 + X^{13}Y^8 + X^{19}Y^{15})}{\prod_{i=0}^{5}(1 - X^iY) \cdot (1 - X^{18}Y^9)(1 - X^8Y^7)(1 - X^{14}Y^8)},$$

$$W_2(X, Y) = \frac{(1 - Y)(1 + Y)X^6Y^5(1 + X^{13}Y^8)}{\prod_{i=0}^{5}(1 - X^iY) \cdot (1 - X^{18}Y^9)(1 - X^8Y^7)(1 - X^7Y^5)(1 - X^{14}Y^8)}.$$

Nonuniformity and the shape of the local zeta functions, i.e. the existence of nonzero rational functions $W_i(X, Y)$ in (7.1.2), were already established by du Sautoy in [3].

Moreover, we showed that in this manner (the number of $\mathbb{F}_p$-points of) every *smooth* plane curve defined over $\mathbb{Q}$ could be realized in local normal zeta functions of nilpotent groups. These improvements became feasible once we interpreted the local zeta function of a nilpotent group as a generating function associated to a weight function defined on the vertex set of the Bruhat-Tits building for $SL_n(\mathbb{Q}_p)$. In [13] we generalized this approach to deal with $\mathfrak{T}$-groups G of nilpotency class 2 – let us call them $\mathfrak{T}_2$-groups – whose Pfaffians $\mathcal{P}_G$ are smooth hypersurfaces containing no lines. This is the generic case for $\mathfrak{T}_2$-groups G with 'small centre and sufficiently large abelianisation', or, more precisely,

$$h(G') \leq 6 \text{ and } h(G/G') > 4h(G') - 10. \qquad (7.1.3)$$

(Here $G' = [G, G]$ is the group's derived group and $h(G)$ denotes the Hirsch length of the polycyclic group G, i.e. the number of infinite cyclic factors in a decomposition series of G.)

In the current paper we treat classes of examples of $\mathfrak{T}_2$-groups whose Pfaffian $\mathcal{P}_G$ is singular. Our main Theorem 7.1.3 illustrates that in general the formulae for the group's local zeta functions may invoke not only (the number of $\mathbb{F}_p$-points of) the Pfaffian hypersurface, but indeed all the strata of its rank stratification.

For the rest of this paper, G shall denote a $\mathfrak{T}_2$-group. Only for simplicity we assume that $G' = Z(G)$, the group's centre, and $G/G' \cong \mathbb{Z}^d$, $d = 2r$, and $G' \cong \mathbb{Z}^{d'}$. We will assume that G is presented as in (7.1.1) above, where $M(\mathbf{y})$ is an anti-symmetric $(d \times d)$-matrix of $\mathbb{Z}$-linear forms in $\mathbf{y} = (y_1, \ldots, y_{d'})$ – but not necessarily in block form as in the particular example (7.1.1). We may then consider $\mathcal{P}_G$ as a hypersurface of degree r in $\mathbb{P}^{d'-1}(\mathbb{Q})$ defined over $\mathbb{Z}$. We say that G is *regular* if

$$\forall' p \text{ prime } \forall \alpha \in \mathbb{P}^{d'-1}(\mathbb{Q}_p): \; rk(\overline{M(\alpha)}) \geq 2(r-1).$$

(Here by $\forall'$ we mean "for almost all", i.e. for all but finitely many.)

Lemma 7.1.2. *If $\mathcal{P}_G \subset \mathbb{P}^{d'-1}(\mathbb{Q})$ is smooth then G is regular.*

Proof. If $\mathcal{P}_G$ is smooth, the reduction $\overline{\mathcal{P}_G}$ is smooth for almost all primes p. If G were not regular there were infinitely many primes p for which there exists a point $\mathbf{x} \in \mathbb{P}^{d'-1}(\mathbb{F}_p)$ for which

$$rk(M(\mathbf{x})) < 2(r-1).$$

It is thus enough to show that for almost all primes such a point must be a singular point of $\overline{\mathcal{P}_G}$.

Let $\bar{i}$ be the reduction mod p of the linear embedding of $\mathbb{P}^{d'-1}(\mathbb{Q})$ into the projective space $\mathbb{P}(\mathcal{S}_r)(\mathbb{Q})$ of anti-symmetric $(2r \times 2r)$-matrices over $\mathbb{Q}$ up to scalars given by

$$i: \mathbb{P}^{d'-1}(\mathbb{Q}) \to \mathbb{P}(\mathcal{S}_r)(\mathbb{Q})$$
$$\alpha \mapsto (\ldots : L_{ij}(\alpha) : \ldots)$$

where the linear forms L_{ij} are given by the ij-th entry of the matrix $M(\alpha)$. Ignoring finitely many primes p as we may we assume that $\bar{i}$ is injective. Let $U := \bar{i}(\mathbb{P}^{d'-1}(\mathbb{F}_p))$ and $\mathcal{X}_r$ denote the universal Pfaffian hypersurface in $\mathbb{P}(\mathcal{S}_r)(\mathbb{Q})$ of *singular* anti-symmetric $(2r \times 2r)$-matrices up to scalars. Its singular locus $\mathrm{Sing}(\mathcal{X}_r)$ consists of matrices of rank strictly less than $2(r-1)$. We have $\bar{i}(\overline{\mathcal{P}_G}) = U \cap \overline{\mathcal{X}_r}$. It is enough to show that

$$U \cap \mathrm{Sing}(\overline{\mathcal{X}_r}) \leq \mathrm{Sing}(\bar{i}(\overline{\mathcal{P}_G})).$$

But that is clear since if $\bar{i}(\mathbf{x}) \in U \cap \mathrm{Sing}(\overline{\mathcal{X}_r})$ then $T_{\bar{i}(\mathbf{x})}(\bar{i}(\overline{\mathcal{P}_G})) = U$. Thus $\mathbf{x}$ is a singular point of $\overline{\mathcal{P}_G}$. □

The converse of Lemma 7.1.2 does not hold, as the following example shows:

Example. Consider the (singular) plane curve

$$C := (y^2 - x^3 + x^2 = 0).$$

We define two groups $G_1(C)$ and $G_2(C)$ as in (7.1.1) above, with

$$R_1(\mathbf{y}) = \begin{pmatrix} y_1 & y_2 & 0 \\ 0 & y_1 & y_2 \\ -y_3 & 0 & y_1 - y_3 \end{pmatrix}, \quad R_2(\mathbf{y}) = \begin{pmatrix} y_1 & y_2 & 0 \\ 0 & y_1 - y_3 & y_2 \\ -y_3 & 0 & y_1 \end{pmatrix}.$$

Note that $C = \mathcal{P}_{G_1(C)} = \mathcal{P}_{G_2(C)}$, but that only $G_2(C)$ is regular. As we will see, this distinction will be picked up by the local normal zeta functions.

To state our main theorem, we have to introduce further notation and hypotheses. Assume for the rest of this chapter that $d' = 3$, that $\sqrt{\det(M(\mathbf{y}))}$ is non-zero and square-free, and that the Pfaffian curve $\mathcal{P}_G$ has irreducible components E_i, $i = 1, \ldots, l$, all of degree > 1, and has at most ordinary double points (ODP's) as singularities, all of which are rational with rational slopes. Given a prime p we write

$$c_{p,i} := |\bar{E}_i|$$

for the number of model points over $\mathbb{F}_p$ of E_i, that is the number of $\mathbb{F}_p$-points of a non-singular curve birationally equivalent to E_i. For $j = 1, 2$ we write

$$n_p^{(j)} := |\{\alpha \in \mathbb{P}^2(\mathbb{F}_p) |\ \alpha \text{ is ODP of } \overline{\mathcal{P}_G} \text{ and } rk(M(\alpha)) = 2(r - j)\}|$$

for the number of $\mathbb{F}_p$-rational ODP's of *rank deficit* $2j$. (In the above example we had $n_p^{(1)} = 0$ and $n_p^{(2)} = 1$ for $G = G_1(C)$, $n_p^{(1)} = 1$ and $n_p^{(2)} = 0$ for $G = G_2(C)$, and $c_{p,1} = 1 + p$ for both groups and all primes p.) With this notation we have

Theorem 7.1.3. *Assume that G is a $\mathfrak{T}_2$-group such that $\sqrt{\det(M(\mathbf{y}))}$ is non-zero and square-free and the curve $\mathcal{P}_G$ has at most rational ordinary double points with rational slopes as singularities and has no lines. Then, for almost all primes p,*

$$\frac{\zeta^{\triangleleft}_{G,p}(s)}{\zeta_{\mathbb{Z}_p^d}(s)\zeta_p((d+3)s - 3d)} = W_1(p, p^{-s}) + \left(\sum_{i=1}^{l} c_{p,i}\right) W_2(p, p^{-s})$$
$$+ n_p^{(1)} W_3(p, p^{-s}) + n_p^{(2)} W_4(p, p^{-s}), \qquad (7.1.4)$$

where

$$W_1(X, Y) = [1 + X^{2r}Y^{2r+1} + X^{2r+1}Y^{2r+1} + X^{4r}Y^{2r+2} + X^{4r+1}Y^{2r+2}$$
$$+ X^{6r+1}Y^{4r+3}]\,(1 - X^{4r+2}Y^{2r+2})^{-1}\,(1 - X^{2r+2}Y^{2r+1})^{-1}, \qquad (7.1.5)$$

$$W_2(X, Y) = \frac{(1 - Y)(1 + Y)X^{2r}Y^{2r-1}(1 + X^{4r+1}Y^{2r+2})}{(1 - X^{2r+1}Y^{2r-1})(1 - X^{2r+2}Y^{2r+1})(1 - X^{4r+2}Y^{2r+2})}, \qquad (7.1.6)$$

$$W_3(X,Y) = \frac{(1-Y)(1+Y)X^{2r}Y^{2r-1}(1-X^{2r+2}Y^{2r-1})(1+X^{4r+1}Y^{2r+2})}{(1-X^{2r+1}Y^{2r-1})^2(1-X^{2r+2}Y^{2r+1})(1-X^{4r+2}Y^{2r+2})}, \tag{7.1.7}$$

$$\begin{aligned} W_4(X,Y) = {} & [(1-Y)^2(1+Y)^2X^{2r}Y^{2r-3}(1+X^{2r+1}Y^{2r-1}) \\ & \times (1+X^{4r+1}Y^{2r+2})]\,(1-X^{2r+2}Y^{2r+1})^{-1}\,(1-X^{2r}Y^{2r-3})^{-1} \\ & \times (1-X^{2r+1}Y^{2r-1})^{-1}\,(1-X^{4r+2}Y^{2r+2})^{-1}. \end{aligned} \tag{7.1.8}$$

In (7.1.4), $\zeta_p(s) = (1-p^{-s})^{-1}$ denotes the p-th local Riemann zeta function. The division on the left hand side is purely formal.

Recall that, by properties of the Weil zeta functions associated to smooth projective varieties defined over finite fields (cf. [1]), for each $i = 1,\dots,l$ there exist integers $B^{(i)}$ and complex numbers $\beta_{i,j}$ such that

$$c_{p,i} = |\overline{E_i}| = 1 + p - \sum_{j=1}^{B^{(i)}} \beta_{i,j}$$

and a one-to-one correspondence

$$\left\{ \frac{p}{\beta_{i,j}} | 1 \le j \le B^{(i)} \right\} \overset{1-1}{\longleftrightarrow} \{\beta_{i,k} | 1 \le k \le B^{(i)}\}.$$

Set

$$c_{p^{-1},i} := 1 + p^{-1} - \sum_{j=1}^{B^{(i)}} \beta_{i,j}^{-1} = p^{-1} c_{p,i}.$$

$$n_{p^{-1}}^{(j)} := n_p^{(j)}.$$

Corollary 7.1.4. *For all but finitely many primes p the local normal zeta function $\zeta^{\triangleleft}_{G,p}(s)$ satisfies a functional equation*

$$\zeta^{\triangleleft}_{G,p}(s)|_{p\to p^{-1}} = -p^{\binom{2r+3}{2} - (4r+3)s}\zeta^{\triangleleft}_{G,p}(s). \tag{7.1.9}$$

Proof (of Corollary 7.1.4). This follows immediately from the observation that

$$W_i(p,p^{-s})|_{p\to p^{-1}} = \begin{cases} p^3 W_i(p,p^{-s}) & \text{for } i = 1,3,4, \\ p^4 W_i(p,p^{-s}) & \text{for } i = 2. \end{cases}$$

□

(Note added in Proof: Very general results establishing local functional equations for zeta functions of rings and nilpotent groups have recently been achieved in [14]. These include results for normal zeta functions of arbitrary $\mathfrak{T}_2$-groups, greatly generalising Corollary 7.1.4.)

Theorem 7.1.1 follows from Theorem 7.1.3 by setting $r = 3, l = 1, c_1 = |E(\mathbb{F}_p)|, n^{(1)} = n^{(2)} = 0$, and that we get explicit formulae for the zeta functions of the groups constructed in the example preceding Theorem 7.1.3.

Before we prove Theorem 7.1.3 in Section 7.3 we give a short introduction to our method in Section 7.2 to make this paper self-contained.

7.2. Counting in Bruhat-Tits buildings

Now let G be again a $\mathfrak{T}_2$-group with

$$Z(G) = G', h(G/G') = d = 2r, h(G') = d'.$$

Nothing will be lost if we assume $d' \geq 2, d \geq 4$. For a fixed prime p, the computation of the p-th normal local zeta function of G comes down to an enumeration of lattices (= subgroups of finite index) in the $\mathbb{Z}_p$-Lie algebra (with Lie brackets induced by taking commutators)

$$G_p := (G/Z(G) \oplus Z(G)) \otimes_{\mathbb{Z}} \mathbb{Z}_p$$

which are ideals in G_p. We call a lattice $\Lambda \leq \mathbb{Z}_p^n$ *maximal* (in its homothety class) if $p^{-1}\Lambda \not\leq \mathbb{Z}_p^n$. The key observation is the following

Lemma 7.2.1. *For each lattice $\Lambda' \leq G'_p$ put $X(\Lambda')/\Lambda' = Z(G_p/\Lambda')$. Then*

$$\begin{aligned}\zeta^{\triangleleft}_{G,p}(s) &= \zeta_{\mathbb{Z}_p^d}(s) \sum_{\Lambda' \leq_f G'_p} |G'_p : \Lambda'|^{d-s} |G_p : X(\Lambda')|^{-s} \\ &= \zeta_{\mathbb{Z}_p^d}(s)\zeta_p((d+d')s - dd') \underbrace{\sum_{\substack{\Lambda' \leq_f G'_p \\ \Lambda' maximal}} |G'_p : \Lambda'|^{d-s} |G_p : X(\Lambda')|^{-s}}_{=:A(p,p^{-s}),say} .\end{aligned}$$

This is essentially a local version of Lemma 6.1 in [8]. One of the main theorems of [8] establishes the rationality of the generating function $A(p, p^{-s})$ in p^{-s}. Note that it is in general not true that $A(p, p^{-s})$ is a rational function in both p^{-s} and p, i.e. that $\zeta^{\triangleleft}_G(s)$ is uniform. Recall that (homothety classes of) maximal lattices are in one-to-one correspondence with vertices of the Bruhat-Tits building $\Delta_{d'}$ for $SL_{d'}(\mathbb{Q}_p)$ (e.g. [6], §19). To derive an explicit formula for $A(p, p^{-s})$ requires a quantitative understanding of two weight functions defined on the set of vertices of the building $\Delta_{d'}$. We write

$$A(p, p^{-s}) = \sum_{[\Lambda']} p^{dw([\Lambda'])-sw'([\Lambda'])}, \tag{7.2.1}$$

where, for a homothety class $[\Lambda']$ of a *maximal* lattice Λ' in $G'_p \cong \mathbb{Z}_p^{d'}$ we define

$$\begin{aligned} w([\Lambda']) &:= \log_p(|G'_p : \Lambda'|), \\ w'([\Lambda']) &:= w([\Lambda']) + \log_p(|G_p : X(\Lambda')|). \end{aligned}$$

In [12] we explain in detail how the evaluation of w' might be reduced to solving linear congruences. To recall the result in Theorem 7.2.2 below we have to introduce some more notation.

A maximal lattice $\Lambda' \leq \mathbb{Z}_p^{d'}$ is said to be of *type* $\nu(\Lambda') = (I, \mathbf{r}_I)$ if Λ' has elementary divisors

$$(p^\nu) := \left(\underbrace{1, \ldots, 1}_{i_1}, \underbrace{p^{r_{i_1}}, \ldots, p^{r_{i_1}}}_{i_2 - i_1}, \ldots, \underbrace{p^{\sum_{j=1}^{l} r_{i_j}}, \ldots, p^{\sum_{j=1}^{l} r_{i_j}}}_{d' - i_l} \right) \tag{7.2.2}$$

for

$$I = \{i_1, \ldots, i_l\} \subseteq [d' - 1], 1 \leq i_1 < \cdots < i_l \leq d' - 1, \tag{7.2.3}$$

and $\mathbf{r}_I = (r_{i_1}, \ldots, r_{i_l}) \in \mathbb{N}_{>0}^l$. Here we set $[n] := \{1, \ldots, n\}$. By abuse of notation we may say that a maximal lattice Λ' is of type I if it is of type $(I, \mathbf{r}_I)$ for some positive vector $\mathbf{r}_I$, and that a (not necessarily maximal) lattice or a homothety class has type I if the maximal element of the class has type I.

The group $\Gamma = SL_n(\mathbb{Z}_p)$ acts transitively on the set of maximal lattices of fixed type. If we choose a basis for the $\mathbb{Z}_p$-module G'_p and represent lattices as the row span of $d' \times d'$-matrices and denote by Γ_ν the stabilizer in Γ of the lattice generated by the diagonal matrix whose entries are given by the vector (7.2.2), the orbit-stabilizer theorem gives us a one-to-one correspondence

$$\{\text{maximal lattices of type } (I, \mathbf{r}_I)\} \stackrel{1-1}{\longleftrightarrow} \Gamma/\Gamma_\nu. \tag{7.2.4}$$

The correspondence (7.2.4) allows us to describe $|G_p : X(\Lambda')|$ for a maximal lattice Λ' in terms of $M(\mathbf{y})$, the matrix of commutators in a presentation for G as in (7.1.1). Its entries are $\mathbb{Z}$-linear forms in generators $y_1, \ldots, y_{d'}$ for G'.

Theorem 7.2.2 ([12], §2.2). *If Λ' corresponds to the coset $\alpha\Gamma_\nu$ under (7.2.4), where $\alpha \in \Gamma$ with column vectors α^j, $j = 1, \ldots, d'$. Then the index $|G_p : X(\Lambda')|$ equals the index of the kernel of the following system of linear congruences in G_p/G'_p:*

$$\forall i \in \{1, \ldots, d'\} \quad \bar{\mathbf{g}} M(\alpha^i) \equiv 0 \; mod \; (p^\nu)_i, \tag{7.2.5}$$

where $\bar{\mathbf{g}} = (\bar{g}_1, \ldots, \bar{g}_d) \in G_p/G'_p \cong \mathbb{Z}_p^d$ and $(p^\nu)_i$ denotes the i-th entry of the vector (p^ν) given in (7.2.2).

Further study of the linear congruences (7.2.5) will enable us to formulate conditions on the group G under which the explicit computation of $A(p, p^{-s})$ – and thus $\zeta^{\triangleleft}_{G,p}(s)$ – is feasible. Given $\alpha \in \Gamma$ and I as in (7.2.3), let $\bar{\alpha}$ denote the reduction mod p and define vector spaces

$$V_{i_j} := \langle \overline{\alpha^{i_j+1}}, \ldots, \overline{\alpha^{d'}} \rangle < \mathbb{P}^{d'-1}(\mathbb{F}_p), \quad i_j \in I.$$

Clearly $codim(V_{i_j}) = i_j$. We will call the nested sequence of subspaces

$$V_{i_1} > \cdots > V_{i_l}$$

the *flag of type I associated to α* or indeed *to the maximal lattice Λ'* if $\nu(\Lambda') = I$ and Λ' corresponds to $\alpha\Gamma_\nu$ under (7.2.4). It is indeed straightforward to show that this is well-defined, i.e. independent of the coset representative α. Given a fixed point $\bar{x} \in \mathbb{P}^{d'-1}(\mathbb{F}_p)$, we say that a lattice Λ' *lifts* $\bar{x}$ if its associated flag contains $\bar{x}$ as 0-dimensional member. Note that then necessarily the type of Λ' contains $d'-1$.

Given $I \leq [d'-1]$ and $\bar{x} \in \mathbb{P}^{d'-1}(\mathbb{F}_p)$, we set

$$A_I(p, p^{-s}) := \sum_{\nu(\Lambda')=I} p^{dw([\Lambda'])-sw'([\Lambda'])},$$

$$A_{\bar{x}}(p, p^{-s}) := \sum_{\Lambda' \text{ lifts } \bar{x}} p^{dw([\Lambda'])-sw'([\Lambda'])}.$$

Clearly

$$A(p, p^{-s}) = \sum_{I \leq [d'-1]} A_I(p, p^{-s}) \tag{7.2.6}$$

$$= \sum_{\bar{x} \in \mathbb{P}^{d'-1}(\mathbb{F}_p)} A_{\bar{x}}(p, p^{-s}) + \sum_{I \not\ni d'-1} A_I(p, p^{-s}) \tag{7.2.7}$$

Formula (7.2.7) gives us an inkling how the combinatorial object $A(p, p^{-s})$ is linked to the geometry of the Pfaffian $\mathcal{P}_G$, a hypersurface in projective $(d'-1)$-space. Theorem 7.2.2 tells us that the number $|G_p : X(\Lambda')|$ depends in the first instance on the way the flag associated to Λ' intersects with $\overline{\mathcal{P}_G}$. Assume from now on that its defining polynomial $\sqrt{\det(M(\mathbf{y}))}$ is non-zero and square-free, and that $\mathcal{P}_G$ contains no lines. This last condition simplifies the system of linear congruences (7.2.5) and is generically satisfied if $r > 2d' - 5$ (see Beauville's Appendix to [13] for details).

In the easiest case $\mathcal{P}_G$ has no $\mathbb{F}_p$-rational points. Thus $\det(M(\alpha))$ is a p-adic unit for all $\alpha \in \mathbb{Z}_p^{d'} \setminus (p\mathbb{Z}_p^{d'})$ and (7.2.5) is equivalent to the single congruence

$$\bar{\mathbf{g}} \equiv 0 \; mod \; p^{\sum_{j=1}^{l} r_{i_j}}.$$

Hence

$$|G_p : X(\Lambda')| = p^{d \sum_{j=1}^{l} r_{i_j}}. \tag{7.2.8}$$

In particular, both weight functions w and w' only depend on the maximal lattices' *type*. Such functions have been studied in [13]:

Theorem 7.2.3 ([13]). *Assume that the weight functions $w([\Lambda'])$ and $w'([\Lambda'])$ only depend on the elementary divisor type of Λ'. Then there exist integers a_i, b_i for $i \in \{1, \ldots, d'-1\}$, depending only on d, d' and i, such that, for all $I \leq [d'-1]$,*

$$A_I(p, p^{-s}) = b_I(p^{-1}) \prod_{i \in I} \frac{X_i}{1 - X_i} \Bigg|_{X_i \to p^{b_i - a_i s}} \tag{7.2.9}$$

where $b_I(p) \in \mathbb{Z}[p]$ denotes the number of $\mathbb{F}_p$-rational points of the projective variety of flags of type I. In particular

$$A_{\bar{x}}(p, p^{-s}) = \left. \underbrace{\frac{p^{-(d'-1)}X_{d'-1}}{1 - X_{d'-1}}}_{=:A^{(1)}(p,p^{-s}),\ say} \left(\sum_{I' \subseteq [d'-2]} b_{I'}(p^{-1}) \prod_{i \in I'} \frac{X_i}{1 - X_i} \right) \right|_{X_i \to p^{b_i - a_i s}}, \tag{7.2.10}$$

independently of $\bar{x} \in \mathbb{P}^{d'-1}(\mathbb{F}_p)$. The following local functional equation holds

$$A(p, p^{-s})|_{p \to p^{-1}} = (-1)^{d'-1} p^{-\binom{d'}{2}} A(p, p^{-s}).$$

In this special case $A(p, p^{-s})$ is an instantiation of a type of functions first studied by Igusa [9]. The polynomials $b_I(p)$ are easily computed in terms of q-binomial polynomials (cf. [11]). The proof of the functional equation relies on a symmetry of the (poset) lattice of Schubert cells for the complete flag variety. We note that this symmetry is independent of the 'numerical data' a_i, b_i. See [13], §4.2.1, for details. There it is also proved that

$$a_i := d + d' - i, \; b_i := (d + i)(d' - i), \; 1 \leq i \leq d' - 1 \tag{7.2.11}$$

is the right numerical data for the case that $\mathcal{P}_G$ has no $\mathbb{F}_p$-points. Formula (7.1.5) may then be retrieved from (7.2.6), (7.2.9) and (7.2.11).

Note that under the assumption that $\mathcal{P}_G$ contains no higher-dimensional linear subspaces, the formulae for the functions $A_I(p, p^{-s})$, $d' - 1 \notin I$, are given by (7.2.9) regardless if $\mathcal{P}_G$ has $\mathbb{F}_p$-rational points, at least for almost all primes p. This follows from inspection of the linear congruences (7.2.5). It can also be shown ([13], Prop. 4) that, under this assumption, for $\bar{x} \in \mathbb{P}^{d'-1}(\mathbb{F}_p)$, for almost all primes, $A_{\bar{x}}(p, p^{-s})$ equals

$$\left. \underbrace{\left(\sum_{\substack{\Lambda' \text{ lifts } \bar{x} \\ \nu(\Lambda') = d'-1}} p^{dw([\Lambda']) - sw'([\Lambda'])} \right)}_{=:A'_{\bar{x}}(p,p^{-s})} \left(\sum_{I' \subseteq [d'-2]} b_{I'}(p^{-1}) \prod_{i \in I'} \frac{X_i}{1 - X_i} \right) \right|_{X_i \to p^{b_i - a_i s}} \tag{7.2.12}$$

If $\bar{x} \notin \overline{\mathcal{P}_G}$ and the numerical data is as in (7.2.11) above, this formula coincides with the one given in (7.2.10). If $\bar{x}$ is a rational point of $\overline{\mathcal{P}_G}$, however, equation (7.2.12) tells us that the mistake we have made by choosing (7.2.10) instead is localized in the first factor. To correct this mistake we have to compute $A'_{\bar{x}}(p, p^{-s})$, a sum which extends just over lattices lifting the (highly incomplete) flag consisting just of the point $\bar{x}$. This generating function will depend on the structure of $\mathcal{P}_G(\mathbb{Q}_p)$ near $\bar{x}$. The easiest case – that $\bar{x}$ is a smooth point of $\overline{\mathcal{P}_G}$ – was dealt with in [13] and will be recalled in the remainder of this chapter. The main point of Theorem 7.1.3 is to illustrate that to deal with even the tamest singularities requires additional information (here on the rank filtration) and intricate computations. A satisfactory machinery to compute $A'_{\bar{x}}(p, p^{-s})$ in general would be highly desirable but seems out of reach by far. It would have to implement – among other things – a resolution of singularities for $\mathcal{P}_G$.

If $\bar{x}$ is a rational point of $\bar{\mathcal{P}_G}$, the linear congruences (7.2.5) to be solved in order to compute $A'_{\bar{x}}(p, p^{-s})$ are

$$\bar{\mathbf{g}} M(\alpha^{d'}) \equiv 0 \mod p^{r_{d'-1}}, \tag{7.2.13}$$

where $\alpha^{d'} \in \mathbb{P}^{d'-1}(\mathbb{Z}/(p^{r_{i_l}}))$ is a lift of $\bar{x}$. If $\bar{x}$ is a smooth point of $\mathcal{P}_G$ we can find local coordinates such that (7.2.13) reads like

$$\bar{\mathbf{g}} \operatorname{diag}\left(\begin{pmatrix} 0 & x \\ -x & 0 \end{pmatrix}, J_2, \ldots, J_2\right) \equiv 0 \mod p^{r_{d'-1}}, \tag{7.2.14}$$

where $J_2 = \left(\begin{smallmatrix} rl0 & 1 \\ -1 & 0 \end{smallmatrix}\right)$ and $x \in p\mathbb{Z}_p/(p^{d'-1})$. Thus

$$|G_p : X(\Lambda')| = p^{d\sum_{j=1}^{l} r_{i_j} - 2\min\{r_{d'-1}, v_p(x)\}}, \tag{7.2.15}$$

with p-adic valuation v_p.

A quantitative version of Hensel's Lemma describes how $v_p(x)$ varies as Λ' runs over the maximal lattices lifting $\bar{x}$ and allows us to compute the generating function

$$A'_{\bar{x}}(p, p^{-s}) = \frac{p^{2r-s(2r-1)}(1 - p^{2r+1-s(2r+1)})}{(1 - p^{2r+1-s(2r-1)})(1 - p^{2r+2-s(2r+1)})} = A^{(2)}(p, p^{-s}), \text{ say,}$$

and thus the "correction term" $W_2(p, p^{-s})$ equals

$$\left(A^{(2)}(p, p^{-s}) - A^{(1)}(p, p^{-s})\right)\left(\sum_{I' \subseteq [d'-2]} b_{I'}(p^{-1}) \prod_{i \in I'} \frac{X_i}{1 - X_i}\right)\Bigg|_{X_i \to p^{b_i - a_i s}},$$

giving us (7.1.6) explicitly. See [13], §4.2, for details. A variant of this Lemma will be derived in Section 7.3 (formula (7.3.3)) to deal with singular points.

7.3. Proof of theorem 7.1.3

To prove Theorem 7.1.3 we have to compute the two 'correction terms' $W_3(p, p^{-s})$ and $W_4(p, p^{-s})$. We have seen that it suffices to compute $A'_{\bar{x}}(p, p^{-s})$ for a fixed point of rank deficit 2 and 4, respectively. Note that $d' = 3$.

7.3.1. $W_4(p, p^{-s})$

Let $\bar{x}$ be a fixed ODP of $\overline{\mathcal{P}_G}$ of rank deficit 4. Dismissing finitely many primes if necessary, we may assume that we can find local affine coordinates such that condition (7.2.13) reads like

$$\bar{\mathbf{g}} \operatorname{diag}\left(\begin{pmatrix} 0 & x_1 \\ -x_1 & 0 \end{pmatrix}, \begin{pmatrix} 0 & x_2 \\ -x_2 & 0 \end{pmatrix}, J_2\right) \equiv 0 \mod p^a, \tag{7.3.1}$$

($x_1, x_2 \in p\mathbb{Z}/(p^a)$, $a = r_2 \geq 1$). We compute the weight function as

$$\boxed{w'([\Lambda']) = (d+1)a - 2\left(\min\{a, v_p(x_1)\} + \min\{a, v_p(x_2)\}\right).}$$

We shall factorize the map

$$[\Lambda'] \mapsto p^{dw([\Lambda'])-sw'([\Lambda'])}$$

over the set $N := \{(a,b,c) \in \mathbb{N}^3_{>0} \mid a \geq b, a \geq c\}$ as $\psi \circ \phi$ where

$$\phi : [\Lambda'] \mapsto (a, v_p(x_1), v_p(x_2))$$
$$\psi : (a,b,c) \mapsto p^{2ra-s((2r+1)a-2\min(a,b)-2\min(a,c))} \tag{7.3.2}$$

To deduce an explicit formula for the rational generating function $A_{\bar{x}}(p, p^{-s})$ we must study the cardinalities of the fibres of ϕ and must brake up the cone N in order to eliminate the "min" in (7.3.2). It is indeed not hard to compute

$$|\phi^{-1}(a,b,c)| = \begin{cases} 1 & \text{if } a = b = c, \\ (1-p^{-1})\, p^{a-b} & \text{if } a > b, a = c, \\ (1-p^{-1})\, p^{a-c} & \text{if } a > c, a = b, \\ (1-p^{-1})^2 p^{2a-b-c} & \text{if } a > b, a > c, \end{cases} \tag{7.3.3}$$

and to see that the following decomposition of N suits our needs:

$$N := N_0 + N_1 + N_2 + N_3, \text{ where} \tag{7.3.4}$$
$$N_0 := \{(a,b,c) \in N \mid a = b = c\},$$
$$N_1 := \{(a,b,c) \in N \mid a = c > b\},$$
$$N_2 := \{(a,b,c) \in N \mid a = b > c\},$$
$$N_3 := \{(a,b,c) \in N \mid a > b,\ a > c\}.$$

Indeed, if we set $n_j := \dim(N_j)$, we have

$$(a,b,c) \in N_j \Rightarrow \frac{|\phi^{-1}(a,b,c)|\psi(a,b,c)}{(1-p^{-1})^{n_j-1}}$$
$$= m_{j,X}(p,p^{-s})^a m_{j,Y}(p,p^{-s})^b m_{j,Z}(p,p^{-s})^c \tag{7.3.5}$$

for suitable Laurent monomials[1] $m_{j,X}(p,p^{-s})$, $m_{j,Y}(p,p^{-s})$ and $m_{j,Z}(p,p^{-s})$ in variables p and p^{-s}.

Let $F_j(X,Y,Z)$, $j = 0, \ldots, 3$ be the zeta function of the cone N_j, i.e. the rational function

$$F_j(X,Y,Z) = \sum_{(a,b,c) \in N_j} X^a Y^b Z^c.$$

[1] A *Laurent monomial* in variables $x_1, \ldots, x_n$ is a power product $x_1^{r_1} \ldots x_n^{r_n}$ with $r_i \in \mathbb{Z}$.

Table 1. Computing $A^{(4)}(p, p^{-s})$.

j	n_j	$F_j(X, Y, Z)$	$m_{j,X}(p, p^{-s})$	$m_{j,Y}(p, p^{-s})$	$m_{j,Z}(p, p^{-s})$
1	1	$\frac{XYZ}{1-XYZ}$	$p^{2r-s(2r+1)}$	p^{-2s}	p^{-2s}
2	2	$\frac{X^2YZ^2}{(1-XYZ)(1-XZ)}$	$p^{2r+1-s(2r+1)}$	p^{-1-2s}	p^{-2s}
3	2	$\frac{X^2Y^2Z}{(1-XYZ)(1-XY)}$	$p^{2r+1-s(2r+1)}$	p^{-2s}	p^{-1-2s}
4	3	$\frac{X^2YZ(1-X^2YZ)}{(1-XYZ)(1-XY)(1-XZ)(1-X)}$	$p^{2r+2-s(2r+1)}$	p^{-1-2s}	p^{-1-2s}

Simply substituting and summing up yields

$$A^{(4)}(p, p^{-s}) := A'_{\bar{x}}(p, p^{-s})$$
$$= \sum_{\substack{\Lambda' \text{ lifts } \bar{x} \\ \nu(\Lambda')=2}} p^{dw([\Lambda'])-sw'([\Lambda'])}$$
$$= \sum_{j=0}^{3} \sum_{(a,b,c)\in N_j} |\phi^{-1}(a, b, c)|\psi(a, b, c)$$
$$= \sum_{j=0}^{3} (1 - p^{-1})^{n_j-1} F_j(X, Y, Z)\Big|_{\substack{X=m_{j,X}(p,p^{-s}) \\ Y=m_{j,Y}(p,p^{-s}) \\ Z=m_{j,Z}(p,p^{-s})}} = p^{2r-s(2r-3)}$$
$$\times \frac{1+p^{2r+1-s(2r-1)} - 2p^{2r-s(2r-1)} - 2p^{2r-1-s(2r+1)} + p^{4r+1-4sr} + p^{2r-s(2r+1)}}{(1-p^{2r-s(2r-3)})(1-p^{2r+1-s(2r-1)})(1-p^{2r+2-s(2r+1)})},$$

Table 1 records the generating functions $F_j(X, Y, Z)$ together with the integers n_j and the Laurent monomials $m_{j,X}(p, p^{-s})$, $m_{j,Y}(p, p^{-s})$, $m_{j,Z}(p, p^{-s})$. The latter are easily read off from (7.3.2) and (7.3.3). Routine computations with rational functions now yield the desired expression (7.1.8) for

$$W_4(p, p^{-s}) = (A^{(4)}(p, p^{-s}) - 2A^{(2)}(p, p^{-s}) + A^{(1)}(p, p^{-s}))$$
$$\times \left(1 + (p^{-1} + 1)\frac{X_1}{1 - X_1}\right)\Big|_{X_1 \to p^{4r+2-s(2r+2)}} \tag{7.3.6}$$

7.3.2. $W_3(p, p^{-s})$

Now let $\bar{x}$ be a fixed ODP of $\overline{\mathcal{P}_G}$ of rank deficit 2. Ignoring finitely many primes as we may, we choose local coordinates such that condition (7.2.13) becomes

$$\bar{\mathbf{g}} \operatorname{diag}\left(\begin{pmatrix} 0 & x_1x_2 \\ -x_1x_2 & 0 \end{pmatrix}, J_2, J_2\right) \equiv 0 \mod p^a, \tag{7.3.7}$$

Table 2. Computing $A^{(3)}(p, p^{-s})$.

j	n_j	$F_j(X, Y, Z)$	$m_{j,X}(p, p^{-s})$	$m_{j,Y}(p, p^{-s})$	$m_{j,Z}(p, p^{-s})$
0	1	$\frac{XYZ}{1-XYZ}$	$p^{2r-s(2r-1)}$	1	1
1	2	$\frac{X^2YZ^2}{(1-XYZ)(1-XZ)}$	$p^{2r+1-s(2r-1)}$	p^{-1}	1
2	2	$\frac{X^2Y^2Z}{(1-XYZ)(1-XY)}$	$p^{2r+1-s(2r-1)}$	1	p^{-1}
3	3	$\frac{X^2YZ}{(1-XY)(1-XZ)}$	$p^{2r+2-s(2r-1)}$	p^{-1}	p^{-1}
4	3	$\frac{X^3YZ}{(1-XY)(1-XZ)(1-X)}$	$p^{2r+2-s(2r+1)}$	p^{-1-2s}	p^{-1-2s}
5	3	$\frac{X^3Y^2Z^2}{(1-XY)(1-XZ)(1-XYZ)}$	$p^{2r+2-s(2r-1)}$	p^{-1}	p^{-1}

($x_1, x_2 \in p\mathbb{Z}/(p^a)$, $a = r_2 \geq 1$). We compute the weight function

$$\boxed{w'([\Lambda']) = (d+1)a - 2\min\{a, v_p(x_1) + v_p(x_2)\}.}$$

Again we shall factorize

$$[\Lambda'] \mapsto p^{dw([\Lambda'])-sw'([\Lambda'])}$$

over the set $N := \{(a, b, c) \in \mathbb{N}_{>0}^3 | \; a \geq b, a \geq c\}$ as $\psi \circ \phi$ where

$$\phi : [\Lambda'] \mapsto (s, v_p(x_1), v_p(x_2))$$
$$\psi : (a, b, c) \mapsto p^{2ra-s(2r+1)a-2\min(a,b+c)}$$

Recall the formula for $|\phi^{-1}(a, b, c)|$ given in (7.3.3). The subdivision (7.3.4) of N, however, has to be refined to

$$N = \sum_{i=0}^{5} N_i, \text{, where } N_0, N_1, N_2 \text{ as in (7.3.4) and} \tag{7.3.8}$$

$$N_3 := \{(a, b, c) \in N | \; a = b + c\},$$
$$N_4 := \{(a, b, c) \in N | \; a > b + c\},$$
$$N_5 := \{(a, b, c) \in N | \; a < b + c\}.$$

If we set $n_j := \dim(N_j)$ for $j = 0, 1, 2$ and $n_j := 3$ for $j = 3, 4, 5$, equation (7.3.5) holds again all j and suitable Laurent monomials, recorded in Table 2.

Substituting and summing up yields

$$A^{(3)}(p, p^{-s}) := A'_{\bar{x}}(p, p^{-s}) = \sum_{\substack{\Lambda' \text{ lifts } \bar{x} \\ v(\Lambda')=2}} p^{dw([\Lambda'])-sw'([\Lambda'])}$$

$$= \sum_{j=0}^{5} \sum_{(a,b,c)\in N_j} |\phi^{-1}(a,b,c)|\psi(a,b,c)$$

$$= \sum_{j=0}^{5} (1-p^{-1})^{n_j-1} F_j(X,Y,Z)\Big|_{\substack{X=m_{jX}(p,p^{-s})\\ Y=m_{jY}(p,p^{-s})\\ Z=m_{jZ}(p,p^{-s})}}$$

$$= \frac{p^{2r-s(2r-1)}(1-2p^{2r+1-s(2r-1)}+p^{2r+2-s(2r-1)}-p^{2r+2-s(2r+1)}+p^{4r+2-s4r})}{(1-p^{2r+1-s(2r-1)})^2(1-p^{2r+2-s(2r+1)})} \tag{7.3.9}$$

and finally the desired formula (7.1.7) for

$$W_3(p,p^{-s}) = \left(A^{(3)}(p,p^{-s}) - 2A^{(2)}(p,p^{-s}) + A^{(1)}(p,p^{-s})\right) \times \left(1+(p^{-1}+1)\frac{X_1}{1-X_1}\right)\Big|_{X_1\to p^{4r+2-s(2r+2)}}. \tag{7.3.10}$$

See Table 2 for the zeta functions of the respective cones and the corresponding Laurent monomials.

Acknowledgement

This article was written when the author held a Postdoctoral Fellowship from the UK's Engineering and Physical Sciences Research Council (EPSRC). It comprises a part of the author's Cambridge PhD-Thesis, supported by the Studienstiftung des deutschen Volkes and the Cambridge European Trust. We would like to thank Marcus du Sautoy for his encouragement as supervisor, Fritz Grunewald for his continuous support and invaluable conversations, Burt Totaro for asking me about singular Pfaffians, and many others for sharing their knowledge and enthusiasm with me.

References

[1] J. Denef and D. Meuser, A functional equation of Igusa's local zeta function, *Amer. J. Math.*, **113** (1991) 1135–1152.

[2] M. P. F. du Sautoy, A nilpotent group and its elliptic curve: non-uniformity of local zeta functions of groups, *Israel J. Math.*, **126** (2001) 269–288.

[3] M. P. F. du Sautoy, Counting subgroups in nilpotent groups and points on elliptic curves, *J. Reine Angew. Math.*, **549** (2002) 1–21.

[4] M. P. F. du Sautoy, Zeta functions of groups: The quest for order versus the flight from ennui, Groups St. Andrews 2001 in Oxford, CUP, Cambridge (2003).

[5] M. P. F. du Sautoy and D. Segal, Zeta functions of groups, New horizons in pro-p groups, Progr. Math., Birkhæuser, Boston MA (2000) pp. 249–286.

[6] P. Garrett, Buildings and classical groups, Chapman & Hall (1997).

[7] F. J. Grunewald and M. P. F. du Sautoy, Analytic properties of zeta functions and subgroup growth, *Ann. Math.*, **152** (2000) 793–833.

[8] F. J. Grunewald, D. Segal and G. C. Smith, Subgroups of finite index in nilpotent groups, *Invent. Math.*, **93** (1988) 185–223.

[9] J.-I. Igusa, Universal p-adic zeta functions and their functional equations, *Amer. J. Math.*, **111** (1989) 671–716.

[10] A. Lubotzky and D. Segal, Subgroup growth, Birkhäuser Verlag (2003).

[11] I. G. Macdonald, Symmetric functions and Hall polynomials, Oxford Mathematical Monographs (1995).

[12] C. Voll, Zeta functions of groups and enumeration in Bruhat-Tits buildings, *Amer. J. Math.*, **126** (2004) 1005–1032.

[13] C. Voll, Functional equations for local normal zeta functions of nilpotent groups, *Geom. Func. Anal. (GAFA)*, **15** (2005) 274–295, with an appendix by A. Beauville.

[14] C. Voll, Functional equations for zeta functions of groups and rings, to appear in Ann. of Math., math.GR/0612511 on arxiv.org.

Made in United States
North Haven, CT
05 November 2026

21791760R00096